Quick Reference for the Mechanical Engineering PE Exam

Third Edition

Michael R. Lindeburg, PE

Professional Publications, Inc. • Belmont, CA

QUICK REFERENCE FOR THE MECHANICAL ENGINEERING PE EXAM
Third Edition

Printed in the United States of America

Professional Publications, Inc.
1250 Fifth Avenue, Belmont, CA 94002
(650) 593-9119
www.ppi2pass.com

Current printing of this edition: 3

Library of Congress Cataloging-in-Publication Data
Lindeburg, Michael R.
Quick reference for the mechanical engineering PE exam / Michael R. Lindeburg. -- 3rd ed.
p. cm.
ISBN 1-888577-14-2 (saddle-stitch)
1. Mechanical engineering--Problems, exercises, etc.
2. Mechanical engineering--Examinations, questions, etc. I. Title.
TJ159.L524 1997

621'.076--dc21 97-7740
CIP

TABLE OF CONTENTS

HOW TO USE THIS BOOK

The "Quick Reference" series was developed to help you solve problems in the PE exam quickly. This book brings together some of the most important concepts, equations, and data that you might need to solve a problem. With this Quick Reference, you won't need to wade through pages of instruction when all you need is a quick look at the formula.

The idea is to study from your Reference Manual and use your Quick Reference for the exam.

Once you've studied and mastered the theory behind a subject, you're ready to tackle practice problems. Use this Quick Reference when you are solving such problems. Get used to the material in this Quick Reference and its organization.

And, you can use all available blank areas to write in information that is important for you. There are some blank pages at the end for this purpose.

PROFESSIONAL PUBLICATIONS, INC. • Belmont, CA

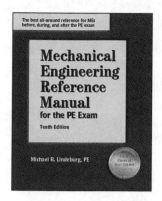

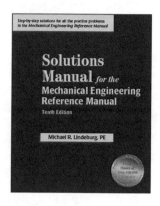

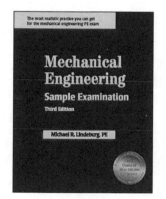

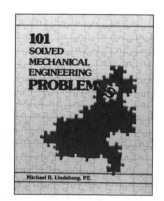

FOR INSTANT RECALL

FUNDAMENTAL CONSTANTS

N_A Avogadro's number: 6.022×10^{23} mol^{-1}

R^* universal gas constant: 1545 ft-lbf/lbmol-°R; 8314 J/kmol·K

c speed of light: 9.84×10^8 ft/sec; 3×10^8 m/s

g_c gravitational constant: 32.174 lbm-ft/lbf-sec^2

J Joule's constant: 778.17 ft-lbf/BTU

TEMPERATURE CONVERSIONS

(Use of the degree symbol with the Kelvin temperature scale is not standard.)

$$°F = 32 + \left(\frac{9}{5}\right) °C$$

$$°C = \left(\frac{5}{9}\right) (°F - 32)$$

$$°R = °F + 460$$

$$°K = °C + 273$$

$$\Delta°R = \Delta°F = \left(\frac{9}{5}\right) \Delta°K = \left(\frac{9}{5}\right) \Delta°C$$

$$\Delta°K = \Delta°C = \left(\frac{5}{9}\right) \Delta°R = \left(\frac{5}{9}\right) \Delta°F$$

DERIVED DATA AND PHYSICAL PROPERTIES

atmospheric pressure: 14.696 psia; 29.92 in of mercury; 407.1 in of water; 1.0133×10^5 Pa

circle: 360°; 2π rad

density of air at 1 atm and 70°F: 0.075 lbm/ft^3; 1.20 kg/m^3

density of mercury: 0.491 lbm/in^3; 1.360×10^4 kg/m^3

density of water: 62.4 lbm/ft^3; 0.0361 lbm/in^3; 1 g/cm^3; 1000 kg/m^3

typical film coefficients:

still air	1.65 BTU/hr-ft^2-°F; 9.37 W/m^2·K
moving air	6.0 BTU/hr-ft^2-°F; 34 W/m^2·K
steam (condensing)	2000 BTU/hr-ft^2-°F; 11,000 W/m^2·K

frequency of house current: 60 Hz; 377 rad/s

standard gravity: 32.17 ft/sec^2; 386 in/sec^2; 9.807 m/s^2

heat of combustion:

hydrogen (high)	60,958 BTU/lbm; 1.419×10^5 kJ/kg
gasoline (high)	20,260 BTU/lbm; 4.716×10^4 kJ/kg
no. 1 diesel (high)	19,240 BTU/lbm; 4.479×10^4 kJ/kg
ethyl alcohol (high)	12,800 BTU/lbm; 2.980×10^4 kJ/kg
carbon	14,093 BTU/lbm; 3.281×10^4 kJ/kg
sulfur	3,983 BTU/lbm; 9.272×10^3 kJ/kg
coal	≈ 12,000 BTU/lbm; 2.8×10^4 kJ/kg

modulus of elasticity for steel: 2.9×10^7 psi; 2×10^{11} Pa

modulus of shear for steel: 1.15×10^7 psi; 7.9×10^{10} Pa

molecular weight:

air	29.0 lbm/lbmole (kg/kmole)
carbon	12.0 lbm/lbmole (kg/kmole)
carbon dioxide	44.0 lbm/lbmole (kg/kmole)
helium	4.0 lbm/lbmole (kg/kmole)
hydrogen	2.0 lbm/lbmole (kg/kmole)
nitrogen	28.0 lbm/lbmole (kg/kmole)
oxygen	32.0 lbm/lbmole (kg/kmole)

Poisson's ratio for steel: 0.3

ratio of specific heats for air: 1.4

solar constant: 429 BTU/ft^2-hr; 1350 W/m^2

specific gas constant for air: 53.3 ft-lbf/lbm-°R; 287 J/kg·K

specific gravity:

water	1.0
mercury	13.6

approximate specific heats:

ice	0.5 BTU/lbm-°F; 2 kJ/kg·K
water	1.0 BTU/lbm-°F; 4.2 kJ/kg·K
steam	0.5 BTU/lbm-°F; 2 kJ/kg·K

triple point of water: 32.02°F, 0.0888 psia; 273.34K, 0.00592 atm

FORMULAS

area of a circle $A = \pi r^2 = \frac{\pi}{4}d^2$

circumference of a circle $p = 2\pi r = \pi d$

area of a triangle $A = \frac{1}{2}bh$

volume of a sphere $V = \frac{4}{3}\pi r^3 = \frac{\pi}{6}d^3$

area moment of inertia for a rectangle

$$I_{centroid} = \frac{1}{12}bh^3$$

$$I_{side} = \frac{1}{3}bh^3$$

area moment of inertia for a circle

$$I_{centroid} = \frac{1}{4}\pi r^4 = \frac{\pi}{64}d^4$$

polar moment of inertia for a circle

$$J = \frac{1}{2}\pi r^4 = \frac{\pi}{32}d^4$$

mass moment of inertia of cylinder

$$J = \frac{1}{2}mr^2 = \frac{1}{8}md^2$$

pressure and head

$$p = \gamma h = \rho h \times \frac{g}{g_c}$$

rotational speed

$$\omega = 2\pi f = 2\pi\left(\frac{rpm}{60}\right)$$

decibels

$$dB = 10 \log_{10}\left(\frac{P_2}{P_1}\right)$$

FUNDAMENTAL AND PHYSICAL CONSTANTS

quantity	symbol	English	SI
Charge			
electron	e		-1.6022×10^{-19} C
proton	p		$+1.6021 \times 10^{-19}$ C
Density			
air [STP]		0.0805 lbm/ft^3	1.29 kg/m^3
air [70°F (20°C), 1 atm]		0.0749 lbm/ft^3	1.20 kg/m^3
earth [mean]		345 lbm/ft^3	5520 kg/m^3
mercury		849 lbm/ft^3	1.360×10^4 kg/m^3
sea water		64.0 lbm/ft^3	1025 kg/m^3
water [mean]		62.4 lbm/ft^3	1000 kg/m^3
Distance [mean]			
earth radius		2.09×10^7 ft	6.370×10^6 m
earth-moon separation		1.26×10^9 ft	3.84×10^8 m
earth-sun separation		4.89×10^{11} ft	1.49×10^{11} m
moon radius		5.71×10^6 ft	1.74×10^6 m
sun radius		2.28×10^9 ft	6.96×10^8 m
first Bohr radius	a_0	1.736×10^{-10} ft	5.292×10^{-11} m
Gravitational Acceleration			
earth [mean]	g	32.174 (32.2) ft/sec^2	9.8067 (9.81) m/s^2
moon [mean]		5.47 ft/sec^2	1.67 m/s^2
Mass			
atomic mass unit	u	3.66×10^{-27} lbm	1.6606×10^{-27} kg
earth		4.11×10^{23} slugs	6.00×10^{24} kg
earth [customary U.S.]		1.32×10^{25} lbm	n.a.
electron [rest]	m_e	2.008×10^{-30} lbm	9.109×10^{-31} kg
moon		1.623×10^{23} lbm	7.36×10^{22} kg
neutron [rest]	m_n	3.693×10^{-27} lbm	1.675×10^{-27} kg
proton [rest]	m_p	3.688×10^{-27} lbm	1.673×10^{-27} kg
sun		4.387×10^{30} lbm	1.99×10^{30} kg
Pressure, atmospheric		14.696 (14.7) lbf/in^2	1.0133×10^5 Pa
Temperature, standard		32°F (492°R)	0°C (273K)
Velocity			
earth escape		3.67×10^4 ft/sec	1.12×10^4 m/s
light [vacuum]	c	9.84×10^8 ft/sec	2.9979 (3.00) $\times 10^8$ m/s
sound [air, STP]	a	1090 ft/sec	331 m/s
[air, 70°F (20°C)]		1130 ft/sec	344 m/s
Volume, molal ideal gas [STP]		359 ft^3/lbmol	22.41 m^3/kmol
Fundamental Constants			
Avogadro's number	N_A		6.022×10^{23} mol^{-1}
Bohr magneton	μ_B		9.2732×10^{-24} J/T
Boltzmann constant	κ	5.65×10^{-24} ft-lbf/°R	1.3807×10^{-23} J/K
Faraday constant	F		96 487 C/mol
gravitational constant	g_c	32.174 lbm-ft/lbf-sec^2	
gravitational constant	G	3.44×10^{-8} ft^4/lbf-sec^4	6.672×10^{-11} $N\cdot m^2/kg^2$
nuclear magneton	μ_N		5.050×10^{-27} J/T
permeability of a vacuum	μ_0		1.2566×10^{-6} N/A^2 (H/m)
permittivity of a vacuum	ϵ_0		8.854×10^{-12} $C^2/N\cdot m^2$ (F/m)
Planck's constant	h		6.6256×10^{-34} J·s
Rydberg constant	R_∞		1.097×10^7 m^{-1}
specific gas constant, air	R	53.3 ft-lbf/lbm-°R	287 J/kg·K
Stefan-Boltzmann constant		1.71×10^{-9} BTU/ft^2-hr-°R^4	5.670×10^{-8} $W/m^2\cdot K^4$
triple point, water		32.02°F, 0.0888 psia	0.01109°C, 0.6123 kPa
universal gas constant	R^*	1545 ft-lbf/lbmol-°R	8314 J/kmol·K
	R^*	1.986 BTU/lbmol-°R	

CONVERSION FACTORS

to convert	into	multiply by
acres	square feet	43,560.0
acres	square miles	1.562×10^{-3}
ampere-hours	coulombs	3600.0
angstrom units	inches	3.937×10^{-9}
angstrom units	microns	1×10^{-4}
astronomical units	kilometers	1.495×10^{8}
atmospheres	centimeters of mercury	76.0
atmospheres	inches of mercury	29.92
BTU	foot-pounds	778
BTU	horsepower-hours	3.931×10^{-4}
BTU	kilowatt-hours	2.928×10^{-4}
BTU/hour	watts	0.2931
calories	BTU	3.9685×10^{-3}
centimeters	kilometers	1×10^{-5}
centimeters	meters	1×10^{-2}
centimeters	millimeters	10.0
centimeters	feet	3.281×10^{-2}
centimeters	inches	0.3937
coulombs	faradays	1.036×10^{-5}
cubic centimeters	cubic inches	0.06102
cubic centimeters	pints (U.S. liquid)	2.113×10^{-3}
cubic feet	cubic meters	0.02832
cubic feet	gallons	7.48
cubic feet/minute	pounds water/minute	62.43
cubic feet/second	gallons/minute	448.83
days	seconds	86,400.0
degrees (angle)	radians	1.745×10^{-2}
degrees/second	revolutions/minute	0.1667
dynes	newtons	1×10^{-5}
ergs	BTU	9.480×10^{-11}
ergs	foot-pounds	7.3670×10^{-8}
ergs	kilowatt-hours	2.778×10^{-14}
faradays/second	amperes	96,500
feet	centimeters	30.48
feet	meters	0.3048
feet	miles (nautical)	1.645×10^{-4}
feet	miles (statute)	1.894×10^{-4}
feet/minute	centimeters/second	0.5080
feet/second	knots	0.5921
feet/second	miles/hour	0.6818
foot-pounds	BTU	1.286×10^{-3}
foot-pounds	kilowatt-hours	3.766×10^{-7}
gallons	cubic feet	0.1337
gallons	liters	3.785
gallons of water	pounds of water	8.3453
gallons/minute	cubic feet/hour	8.0208
gallons/minute	cubic feet/second	0.00223
grams	ounces (avoirdupois)	3.527×10^{-2}
grams	ounces (troy)	3.215×10^{-2}
grams	pounds	2.205×10^{-3}
horsepower	BTU/minute	42.44
horsepower	foot-pounds/minute	33,000
horsepower	foot-pounds/second	550
horsepower	kilowatts	0.7457
horsepower	watts	745.7
hours	days	4.167×10^{-2}
hours	weeks	5.952×10^{-3}
inches	centimeters	2.540
inches	miles	1.578×10^{-5}
joules	BTU	9.480×10^{-4}
joules	ergs	1×10^{7}
kilograms	pounds	2.205
kilograms	slugs	0.068522
kilometers	feet	3281.0
kilometers	meters	1000.0
kilometers	miles	0.6214
kilometers/hour	knots	0.5396
kilopascals	pounds/square inches	0.145
kilowatts	foot-pounds/second	737.6
kilowatts	horsepower	1.341
kilowatt-hours	BTU	3413.0
knots	feet/hour	6080.0
knots	nautical miles/hour	1.0
knots	statute miles/hour	1.151
light years	miles	5.9×10^{12}
links (surveyor's)	inches	7.92
liters	cubic centimeters	1000.0
liters	cubic inches	61.02
liters	gallons (U.S. liquid)	0.2642
liters	milliliters	1000.0
liters	pints (U.S. liquid)	2.113
meters	centimeters	100.0
meters	feet	3.281
meters	kilometers	1×10^{-3}
meters	miles (nautical)	5.396×10^{-4}
meters	miles (statute)	6.214×10^{-4}
meters	millimeters	1000.0
microns	meters	1×10^{-6}
miles (nautical)	feet	6080.27
miles (statute)	feet	5280.0
miles (nautical)	kilometers	1.853
miles (statute)	kilometers	1.609
miles (nautical)	miles (statute)	1.1516
miles (statute)	miles (nautical)	0.8684
miles/hour	feet/minute	88.0
miles/hour	feet/second	1.467
milligram/liter	parts/million	1.0
milliliters	liters	1×10^{-3}
millimeters	inches	3.937×10^{-2}
newtons	dynes	1×10^{5}
newtons	pounds	0.2248
ounces	pounds	6.25×10^{-2}
ounces (troy)	ounces (avoirdupois)	1.09714
parsecs	miles	1.92×10^{13}
parsecs	kilometers	3.084×10^{13}
pints (liquid)	cubic centimeters	473.2
pints (liquid)	cubic inches	28.87
pints (liquid)	gallons	0.125
pints (liquid)	quarts (liquid)	0.5
pounds	kilograms	0.4536
pounds	ounces	16.0
pounds	ounces (troy)	14.5833
pounds	pounds (troy)	1.21528
quarts (dry)	cubic inches	67.20
quarts (liquid)	cubic inches	57.75
quarts (liquid)	gallons	0.25
quarts (liquid)	liters	0.9463
radians	degrees	57.30
radians	minutes	3438.0
revolutions	degrees	360.0
revolutions/minute	degrees/second	6.0
seconds	minutes	1.667×10^{-2}
slugs	pounds	32.17
tons (long)	kilograms	1016.0
tons (short)	kilograms	907.18
tons (long)	pounds	2240.0
tons (short)	pounds	2000.0
tons (long)	tons (short)	1.120
tons (short)	tons (long)	0.89287
watts	BTU/hour	3.4129
watts	horsepower	1.341×10^{-3}
yards	meters	0.9144
yards	miles (nautical)	4.934×10^{-4}
yards	miles (statute)	5.682×10^{-4}

SI SYSTEM

SELECTED CONVERSION FACTORS TO SI UNITS

	SI Symbol	Multiplier to Convert From Existing Unit to SI Unit	Multiplier to Convert From SI Unit to Existing Unit
Area			
Circular Mil	μm²	506.7	0.001 974
Foot Squared	m²	0.092 9	10.764
Mile Squared	km²	2.590	0.386 1
Yard Squared	m²	0.836 1	1.196
Energy			
Btu (International)	kJ	1.055 1	0.947 8
Erg	μJ	0.1	10.0
Foot Pound-Force	J	1.355 8	0.737 6
Horsepower Hour	MJ	2.684 5	0.372 5
Kilowatt Hour	MJ	3.6	0.277 8
Meter Kilogram-Force	J	9.806 7	0.101 97
Therm	MJ	105.506	0.009 478
Kilogram Calorie (International)	kJ	4.186 8	0.238 8
Force			
Dyne	μN	10.	0.1
Kilogram-Force	N	9.806 7	0.101 97
Ounce-Force	N	0.278 0	3.597
Pound-Force	N	4.448 2	0.224 8
KIP	N	4 448.2	0.000 224 8
Heat			
Btu Per Hour	W	0.293 1	3.412 1
Btu Per (Square Foot Hour)	W/m²	3.154 6	0.317 0
Btu Per (Square Foot Hour °F)	W/(m²·°C)	5.678 3	0.176 1
Btu Inch Per (Square Foot Hour °F)	W/(m·°C)	0.144 2	6.933
Btu Per (Cubic Foot °F)	MJ/(m³·°C)	0.067 1	14.911
Btu Per (Pound °F)	J/(kg·°C)	4 186.8	0.000 238 8
Btu Per Cubic Foot	MJ/m³	0.037 3	26.839
Btu Per Pound	J/kg	2 326.	0.000 430
Length			
Angstrom	nm	0.1	10.0
Foot	m	0.304 8	3.280 8
Inch	mm	25.4	0.039 4
Mil	mm	0.025 4	39.370
Mile	km	1.609 3	0.621 4
Mile (International Nautical)	km	1.852	0.540
Micron	μm	1.0	1.0
Yard	m	0.914 4	1.093 6
Mass (weight)			
Grain	mg	64.799	0.015 4
Ounce (Avoirdupois)	g	28.350	0.035 3
Ounce (Troy)	g	31.103 5	0.032 15
Ton (short 2000 lb.)	kg	907.185	0.001 102
Ton (long 2240 lb.)	kg	1 016.047	0.000 984 2
Slug	kg	14.593 9	0.068 522
Pressure			
Bar	kPa	100.0	0.01
Inch of Water Column (20°C)	kPa	0.248 6	4.021 9
Inch of Mercury (20°C)	kPa	3.374 1	0.296 4
Kilogram-force per Centimeter Squared	kPa	98.067	0.010 2
Millimeters of Mercury (mm·Hg) (20°C)	kPa	0.132 84	7.528
Pounds Per Square Inch (P.S.I.)	kPa	6.894 8	0.145 0
Standard Atmosphere (760 torr)	kPa	101.325	0.009 869
Torr	kPa	0.133 32	7.500 6

SI SYSTEM

SELECTED CONVERSION FACTORS TO SI UNITS (continued)

	SI Symbol	Multiplier to Convert From Existing Unit to SI Unit	Multiplier to Convert From SI Unit to Existing Unit
Power			
Btu (International) Per Hour	W	0.293 1	3.412 2
Foot Pound-Force Per Second	W	1.355 8	0.737 6
Horsepower	kW	0.745 7	1.341
Meter Kilogram-Force Per Second	W	9.806 7	0.101 97
Tons of Refrigeration	kW	3.517	0.284 3
Torque			
Kilogram-Force Meter (kg·m)	N·m	9.806 7	0.101.97
Pound-Force Foot	N·m	1.355 8	0.737 6
Pound-Force Inch	N·m	0.113 0	8.849 5
Gram-Force Centimeter	mN·m	0.098 067	10.197
Temperature			
Fahrenheit	°C	5/9 (°F − 32)	(9/5°C) + 32
Rankine	K	(°F + 459.67)5/9	(°C + 273.16)9/5
Velocity			
Foot Per Second	m/s	0.304 8	3.280 8
Mile Per Hour	m/s	0.447 04	2.236 9
	or	*or*	*or*
	km/h	1.609. 34	0.621 4
Viscosity			
Centipoise	mPa·s	1.0	1.0
Centistoke	μm²/s	1.0	1.0
Volume (Capacity)			
Cubic Foot	*l* (dm³)	28.316 8	0.035 31
Cubic Inch	cm³	16.387 1	0.061 02
Cubic Yard	m³	0.764 6	1.308
Gallon (U.S.)	*l*	3.785	0.264 2
Ounce (U.S. Fluid)	ml	29.574	0.033 8
Pint (U.S. Fluid)	*l*	0.473 2	2.113
Quart (U.S. Fluid)	*l*	0.946 4	1.056 7
Volume Flow (Gas-Air)			
Standard Cubic Foot Per Minute	m³/s	0.000 471 9	2119.
	or	*or*	*or*
	l/s	0.471 9	2.119
	or	*or*	*or*
	ml/s	471.947	0.002 119
Standard Cubic Foot Per Hour	ml/s	7.865 8	0.127 133
	or	*or*	*or*
	μ*l*/s	7 866.	0.000 127
Volume Liquid Flow			
Gallons Per Hour (U.S.)	*l*/s	0.001 052	951.02
Gallons Per Minute (U.S.)	*l*/s	0.063 09	15.850

MATHEMATICS

VOLUME AND SURFACE AREAS

A = cross sectional area or end area
S = lateral surface area
V = enclosed volume

RIGHT CIRCULAR CYLINDER

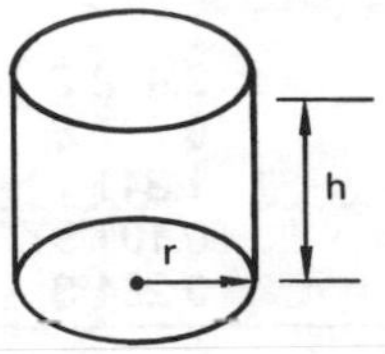

$$A = \pi r^2$$
$$S = 2\pi rh$$
$$V = Ah = \pi r^2 h$$

SPHERE

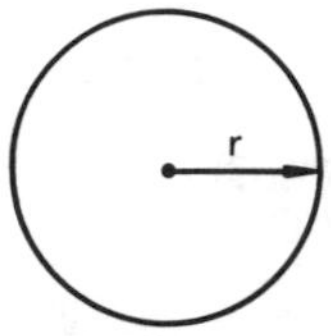

$$S = 4\pi r^2$$
$$V = \left(\tfrac{4}{3}\right)\pi r^3$$

RIGHT CIRCULAR CONE

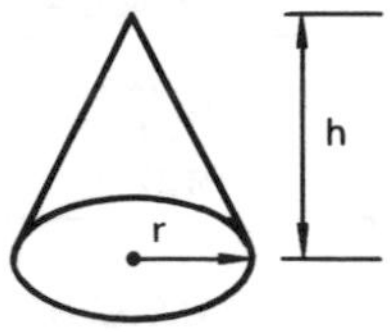

$$A_{base} = \pi r^2$$
$$S = \pi r\sqrt{r^2 + h^2}$$
$$V = \left(\tfrac{1}{3}\right)\pi r^2 h$$

MENSURATION

CIRCLE of radius *r*:

$$\text{circumference} = 2\pi r = p$$
$$\text{area} = \pi r^2 = \frac{p^2}{4\pi}$$

ELLIPSE with axes *a* and *b*

$$\text{area} = \pi ab$$

CIRCULAR SECTOR with radius *r*, included angle θ, and arc length *s*:

$$\text{area} = \tfrac{1}{2}\theta r^2 = \tfrac{1}{2} sr$$
$$\text{arc length} = s = \theta r$$

The angle θ is in radians.

QUADRATIC EQUATION

The roots of the equation $ax^2 + bx + c = 0$ are

$$\frac{-b \pm \sqrt{b^2 - 4ac}}{2a}$$

EXPONENTIATION

$$x^m x^n = x^{(n+m)}$$
$$\frac{x^m}{x^n} = x^{(m-n)}$$
$$(x^n)^m = x^{(mn)}$$
$$a^{m/n} = \sqrt[n]{a^m}$$
$$\left(\frac{a}{b}\right)^n = \frac{a^n}{b^n}$$
$$\sqrt[n]{x} = (x)^{1/n}$$
$$x^{-n} = \frac{1}{x^n}$$
$$x^0 = 1$$

LOGARITHM IDENTITIES

$$x^a = \text{antilog}\,[a \log(x)]$$
$$\log(x^a) = a \log(x)$$
$$\log(xy) = \log(x) + \log(y)$$
$$\log\left(\frac{x}{y}\right) = \log(x)\quad\log(y)$$
$$ln(x) = \frac{\log_{10} x}{\log_{10} e}$$
$$\approx 2.3\,(\log_{10} x)$$
$$\log_b(b) = 1$$
$$\log(1) = 0$$
$$\log_b(b^n) = n$$

TRIGONOMETRY

$$\sin\theta = \frac{y}{h} \qquad \cos\theta = \frac{x}{h} \qquad \tan\theta = \frac{y}{x}$$
$$\csc\theta = \frac{h}{y} \qquad \sec\theta = \frac{h}{x} \qquad \cot\theta = \frac{x}{y}$$

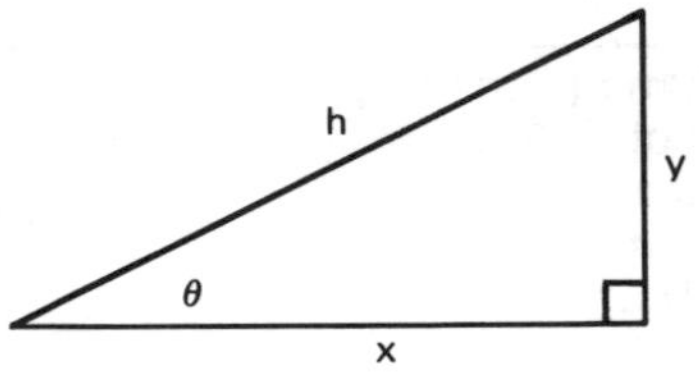

Basic Identities: $\sin^2\theta + \cos^2\theta = 1$

$$\sin 2\theta = 2\sin\theta\cos\theta$$
$$1 + \tan^2\theta = \sec^2\theta$$
$$\cos 2\theta = \cos^2\theta - \sin^2\theta = 2\cos^2\theta - 1$$

For general triangles: the law of sines is

$$\frac{\sin A}{a} = \frac{\sin B}{b} = \frac{\sin C}{c}$$

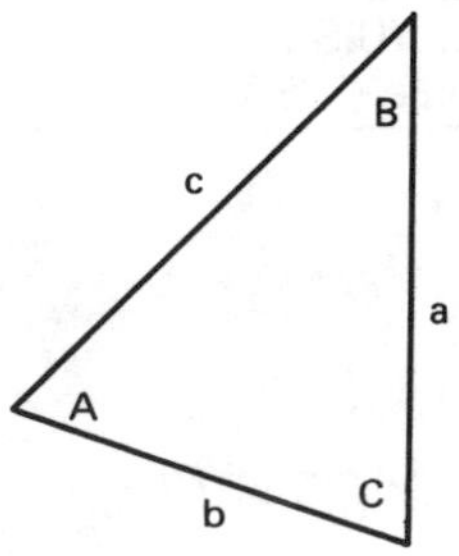

The law of cosines is

$$a^2 = b^2 + c^2 - 2bc\cos A$$

The area of a general triangle is ½ ab(sinC).

EQUATION OF A STRAIGHT LINE

The general form is $Ax + By + C = 0$.

The slope-intercept form is $y = mx + b$, where m is the slope and b is the y-intercept.

Given one point on the line (x^*, y^*), the point-slope form is

$$y - y^* = m(x - x^*).$$

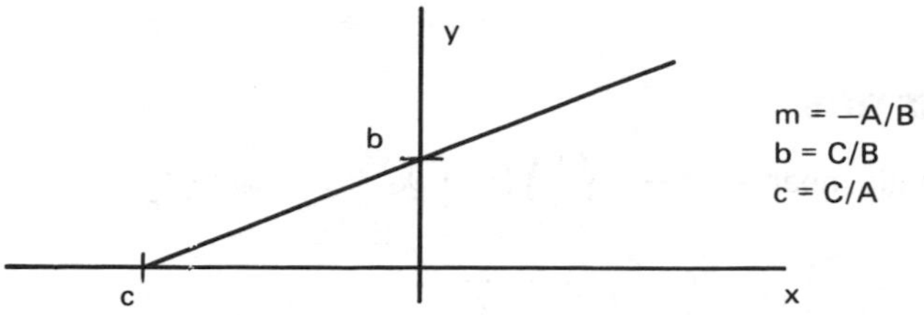

CONIC SECTIONS

The equation of a circle of radius r which is centered at (h,k) is

$$(x-h)^2 + (y-k)^2 = r^2.$$

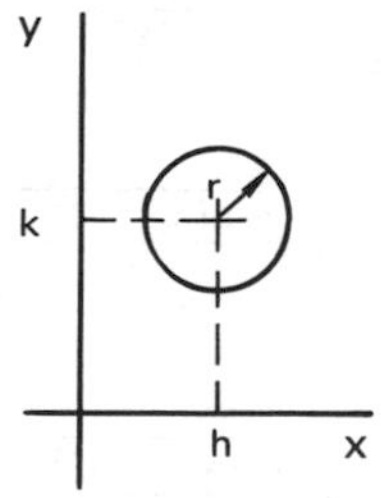

The equation of an ellipse centered at (h,k) with a and b defined as shown is

$$\left(\frac{x-h}{a}\right)^2 + \left(\frac{y-k}{b}\right)^2 = 1$$

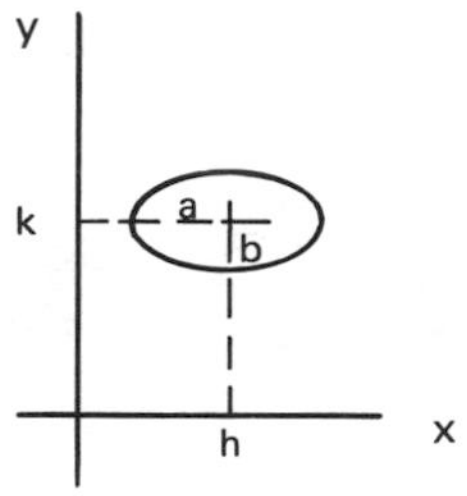

The equation of a parabola with vertex at (h,k) and symmetrical with respect to the y-axis is

$$(x-h)^2 = 4p\,(y-k)$$

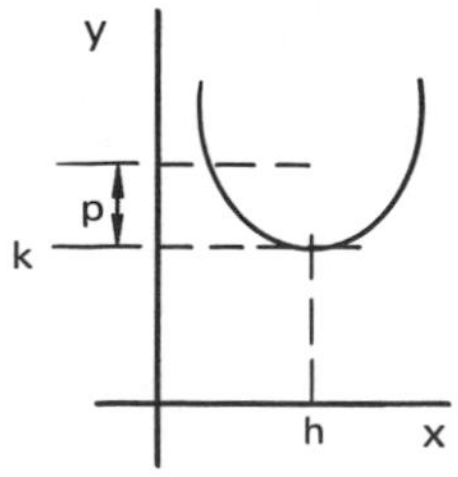

LINEAR REGRESSION

If it is necessary to draw a straight line through n data points $(x_1, y_1), (x_2, y_2), \ldots, (x_n, y_n)$, the following method based on the theory of least squares can be used:

step 1: Calculate the following quantities.

$$\Sigma x_i \quad \Sigma x_i^2 \quad (\Sigma x_i)^2 \quad \bar{x} = \left(\frac{\Sigma x_i}{n}\right) \quad \Sigma x_i y_i$$

$$\Sigma y_i \quad \Sigma y_i^2 \quad (\Sigma y_i)^2 \quad \bar{y} = \left(\frac{\Sigma y_i}{n}\right)$$

step 2: Calculate the slope of the line $y = mx + b$.

$$m = \frac{n\Sigma(x_i y_i) - (\Sigma x_i)(\Sigma y_i)}{n\Sigma x_i^2 - (\Sigma x_i)^2}$$

step 3: Calculate the y intercept.

$$b = \bar{y} - m\bar{x}$$

step 4: To determine the goodness of fit, calculate the correlation coefficient.

$$r = \frac{n\Sigma\,(x_i y_i) - (\Sigma x_i)(\Sigma y_i)}{\sqrt{[n\Sigma x_i^2 - (\Sigma x_i)^2]\,[n\Sigma y_i^2 - (\Sigma y_i)^2]}}$$

EXTREMA BY DIFFERENTIATION

Given a continuous function, f(x), the extreme points may be found by taking the first derivative and setting it equal to zero. Let x^* be the value of x which satisfies this equality. $f(x^*)$ is a minimum if $f''(x^*)$ is greater than zero. If $f''(x^*)$ is less than zero, $f(x^*)$ is a maximum. $f''(x^*)$ is equal to zero at an inflection point.

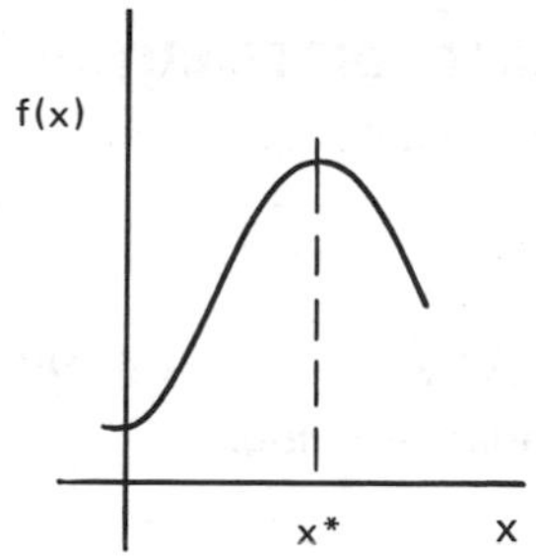

VECTORS

A vector is a directed line segment. It is defined completely only when both the magnitude and direction are known. Vectors are defined in terms of the unit vectors **i,j, k.** The unit vectors are vectors of length one directed along the x, y, and z axes respectively. A vector, **A**, can be written in terms of the unit vectors and its endpoints (a_x, a_y, a_z).

$$\mathbf{A} = a_x\mathbf{i} + a_y\mathbf{j} + a_z\mathbf{k}$$

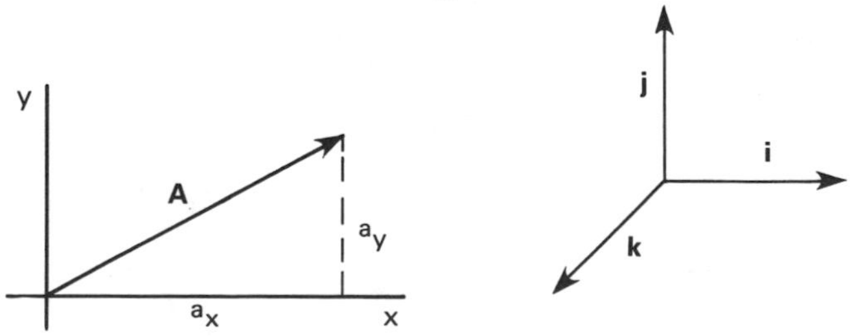

Addition of vectors is performed by adding the components:

$$\mathbf{A} + \mathbf{B} = (a_x + b_x)\mathbf{i} + (a_y + b_y)\mathbf{j} + (a_z + b_z)\mathbf{k}$$

The DOT PRODUCT is a scalar and represents the projection of **A** onto **B**.

$$\mathbf{A}\cdot\mathbf{B} = a_x b_x + a_y b_y = |\mathbf{A}||\mathbf{B}|\cos\theta$$

MATHEMATICS

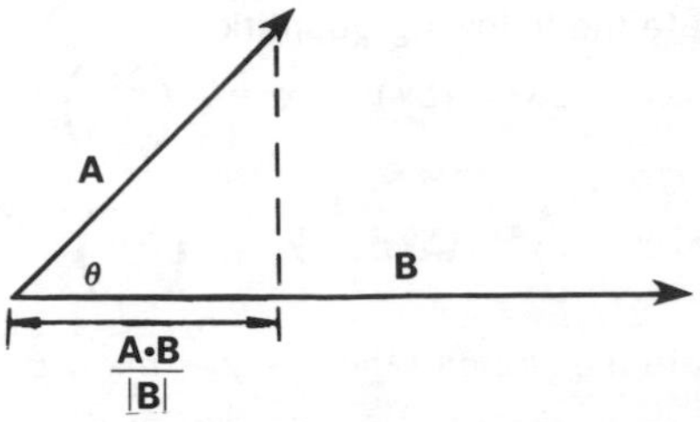

The CROSS PRODUCT is a vector of magnitude |B| |A|sinθ which is perpendicular to the plane containing A and B.

$$\mathbf{A} \times \mathbf{B} = \begin{vmatrix} \mathbf{i} & \mathbf{j} & \mathbf{k} \\ a_x & a_y & a_z \\ b_x & b_y & b_z \end{vmatrix}$$

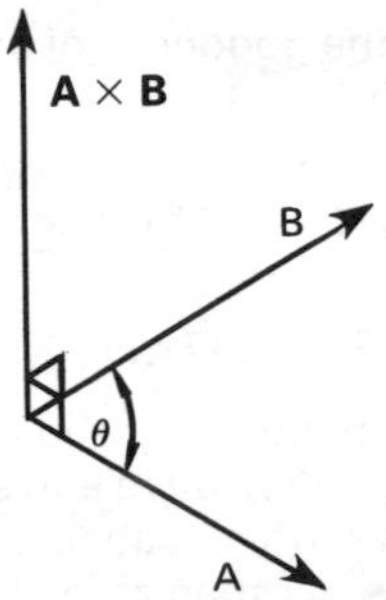

The sense of A×B is determined by the righthand rule.

BINOMIAL PROBABILITY DISTRIBUTION

f(x) is the probability that *x* will occur in *n* trials. *p* is the probability of one success in one trial. q = (1−p) is the probability of one failure in one trial.

$$f(x) = \binom{n}{x} p^x q^{n-x} = \frac{n!}{(n-x)!x!} p^x q^{n-x}$$

The mean is *np*. The variance is *npq*.

POISSON PROBABILITY DISTRIBUTION

f(x) is the probability of *x* occurrences. *x* is the actual number of occurrences in some period. λ is the mean number of occurrences per period.

$$f(x) = \frac{e^{-\lambda}\lambda^x}{x!}$$

The mean and variance are both λ.

STATISTICS

$$\text{arithmetic mean} = \bar{x} = \left(\frac{1}{n}\right)(x_1 + x_2 + \ldots x_n) = \frac{\Sigma x_i}{n}$$

$$\text{geometric mean} = \sqrt[n]{x_1\, x_2\, x_3 \ldots x_n}$$

$$\text{harmonic mean} = \frac{n}{\frac{1}{x_1} + \frac{1}{x_2} + \ldots + \frac{1}{x_n}}$$

$$\text{root-mean-squared value} = \sqrt{\frac{\Sigma x_i^2}{n}}$$

$$\text{standard deviation} = \sigma = \sqrt{\frac{\Sigma(x_i - \bar{x})^2}{n}} = \sqrt{\frac{\Sigma x_i^2}{n} - (\bar{x})^2}$$

$$\text{sample standard deviation} = s = \sqrt{\frac{\Sigma(x_i - \bar{x})^2}{n-1}} = \sqrt{\frac{\Sigma x_i^2 - \frac{(\Sigma x_i)^2}{n}}{n-1}}$$

population variance = σ^2

sample variance = s^2

NOMENCLATURE AND DEFINITIONS

A	annual amount or annuity: a payment or receipt of a fixed sum of money at yearly intervals.
C	cost: the asset purchase price.
d	declining balance depreciation rate: For double-declining balance depreciation, $d = \left(\frac{2}{n}\right)$.
D_j	depreciation in year j.
F	future worth, value, or amount: the value of an asset at some future point in time.
G	uniform gradient amount: a quantity by which the cash flows change each year.
i	annual effective interest rate.
k	number of compounding periods per year.
MARR	Minimum Attractive Rate of Return. Usually the same as the annual effective interest rate, *i*.
n	number of compounding periods; or the expected life of an asset.
P	present worth, value, or amount. The value of an amount at time = 0.
r	nominal annual interest rate. Same as rate per annum.
S_n	expected salvage value in year *n*.
t	time or income tax rate.
ϕ	effective interest rate per period. Equal to $\left(\frac{r}{k}\right)$.

YEAR-END CONVENTION

All cash disbursements and receipts (cash flows) are assumed to occur at the end of the year in which they actually occur. The exception is an initial cost (purchase price) which is assumed to occur at time = 0.

SUNK COSTS

Sunk costs are expenses incurred before time = 0. They have no bearing on the evaluation of alternatives.

CASH FLOW DIAGRAMS

Cash flow diagrams are drawn to help visualize problems involving transfers of money at various points in time. The following conventions are used in their construction:

- The horizontal axis represents time. The axis is divided into equal increments, one per period.
- The year-end convention is assumed.
- Disbursements are downward arrows; receipts are upward arrows. In both cases, the length of the arrow is proportional to the magnitude of the transfer.
- Two or more transfers in the same time period are represented by end-to-end arrows.

DISCOUNTING FACTORS

Discounting factors are numbers used to calculate the equivalent amount (at some point in time) of an alternative. The factors are given the functional notation (X/Y,i%,n) where *X* is the desired value, *Y* is the known value, *i* is the interest rate, and *n* is the number of periods. (Usually, *n* is in years.)

Single Payment Compound Amount	(F/P,i%,n)	$(1+i)^n$
Present Worth	(P/F,i%,n)	$(1+i)^{-n}$
Uniform Series Sinking Fund	(A/F,i%,n)	$\frac{i}{(1+i)^n - 1}$
Capital Recovery	(A/P,i%,n)	$\frac{i(1+i)^n}{(1+i)^n - 1}$
Compound Amount	(F/A,i%,n)	$\frac{(1+i)^n - 1}{i}$
Equal Series Present Worth	(P/A,i%,n)	$\frac{(1+i)^n - 1}{i(1+i)^n}$
Uniform Gradient	(P/G,i%,n)	$\left[\frac{1}{i} - \frac{n}{(1+i)^n - 1}\right]$ (P/A,i%,n)

COMPARING ALTERNATIVES

The PRESENT WORTH method may be used if all of the alternatives have equal lives. The present worth of each alternative is calculated. The alternative with the smallest negative or largest positive present worth is chosen.

The EQUIVALENT UNIFORM ANNUAL COST (EUAC) method must be used if alternatives have unequal lives. Use of this method is restricted to alternatives which are infinitely renewed. Specifically, each alternative is replaced by an identical replacement at the end of its useful life, up to the duration of the longest-lived alternative. The EUAC of an alternative is usually found from the following formula:

$$\text{EUAC} = (\text{present worth of alternative})\,(A/P,i\%,n)$$

The CAPITALIZED COST of a project is the present worth of a project which has an infinite life. The capitalized cost represents the amount of money needed at time = 0 to support the project on interest only, without reducing the principal.

$$\text{capitalized cost} = \text{initial cost} + \frac{(\text{annual maintenance cost})}{i}$$

TREATMENT OF SALVAGE VALUE IN REPLACEMENT STUDIES

By convention, the salvage value is subtracted from the defender's present value. This is done to keep all costs and benefits related to the defender with the defender. In this case, the salvage value is treated as an opportunity cost which would be incurred if the defender is not retired.

RATE OF RETURN

The rate of return (ROR) is the interest rate which makes the present worth equal to zero. To determine the ROR, assume a reasonable value for the interest rate earned and find the present worth. If the present worth is zero, the assumed interest rate is the ROR. If the present worth is not zero, assume another value of *i* and find the present worth. Use the two values of *i* and their corresponding present worths to interpolate or extrapolate a value of *i* which makes the present worth zero.

NON-ANNUAL COMPOUNDING

For problems in which compounding is done at intervals other than yearly, an effective annual interest rate can be computed.

$$i = \left[1 + \left(\frac{r}{k}\right)\right]^k - 1 = (1+\phi)^k - 1$$

DEPRECIATION

STRAIGHT LINE	$D_j = \frac{C - S_n}{n}$
DOUBLE-DECLINING BALANCE (i = 1 to j−1)	$D_j = \frac{2(C - \Sigma D_i)}{n}$
SUM OF THE YEAR'S DIGITS	$D_j = \frac{(C - S_n)(n - j + 1)}{T}$

$$T = \tfrac{1}{2}n(n+1)$$

ENGINEERING ECONOMICS

SINKING FUND $D_j = (C - S_n)\,(A/F,i\%,n)\,(F/P,i,j-1)$

ACRS/MACRS $D_j = (\text{factor})\,C$

INCOME TAXES

If income taxes are paid, operating expenses and depreciation are deductible. If t is the tax rate, revenues and all expenses (except depreciation) should be multiplied by (1–t) in the year in which they occur. Although depreciation is a deductible expense, it is not an actual out-of-pocket expense. Depreciation should be multiplied by t and added to the cash flow in the appropriate year.

CONSUMER LOANS

BAL_j	balance after the jth payment
LV	total value loaned (cost minus down payment)
j	payment or period number
N	total number of payments to pay off the loan
PI_j	jth interest payment
PP_j	jth principal payment
PT_j	jth total payment
ϕ	effective rate per period $\left(\frac{r}{k}\right)$

SIMPLE INTEREST

Interest due does not compound with a *simple interest* loan. The interest due is proportional to the length of time the principal is outstanding.

DIRECT REDUCTION LOANS

This is the typical 'interest paid on unpaid balance' loan. The amount of the periodic payment is constant, but the amounts paid towards the principal and interest both vary.

$$N = -\frac{\ln\left(\frac{-\phi(LV)}{PT} + 1\right)}{\ln(1+\phi)}$$

$$PT = LV\,(A/P,\,i\%,\,n)$$

$$LV = \frac{PT}{(A/P,\,i\%,\,n)}$$

$$BAL_{j-1} = PT\left[\frac{1-(1+\phi)^{j-1-N}}{\phi}\right]$$

$$PI_j = \phi(BAL_j)$$

$$PP_j = PT - PI_j$$

$$BAL_j = BAL_{j-1} - PP_j$$

HANDLING INFLATION

It is important to perform economic studies in terms of *constant value dollars*. One method of converting all cash flows to constant value dollars is to divide the flows by some annual *economic indicator* or price index.

An alternative is to replace i with a value corrected for the inflation rate, e. This corrected value, i', is

$$i' = i + e + ie$$

IMPORTANT FLUID CONVERSIONS

multiply	by	to get
cubic feet	7.4805	gallons
cfs	448.83	gpm
cfs	0.64632	MGD
gallons	0.1337	cubic feet
gpm	0.002228	cfs
inches of mercury	0.491	psi
inches of mercury	70.7	psf
inches of mercury	13.60	inches of water
inches of water	5.199	psf
inches of water	0.0361	psi
inches of water	0.0735	inches of mercury
psi	144	psf
psi	2.308	feet of water
psi	27.7	inches of water
psi	2.037	inches of mercury
psf	0.006944	psi

DENSITY

density of water: 62.4 lbm/ft^3, 0.0361 lbm/in^3; 1000 kg/m^3

density of mercury: 848.4 lbm/ft^3, 0.491 lbm/in^3; 1.36×10^4 kg/m^3

SPECIFIC WEIGHT

$$\gamma = \rho g$$

VISCOSITY

μ: absolute viscosity $\left(\frac{\text{lbf-sec}}{\text{ft}^2}\right)$

ν: kinematic viscosity $\left(\frac{\text{ft}^2}{\text{sec}}\right) = \frac{\mu g_c}{\rho}$

TYPICAL VISCOSITY UNITS

	Absolute	Kinematic
English	lbf-sec/ft² (slug/ft-sec)	ft²/sec
Conventional Metric	dyne-sec/cm² (poise)	cm²/sec (stoke)
SI	Pascal-second (N-s/m²)	m²/s

HYDROSTATIC PRESSURE

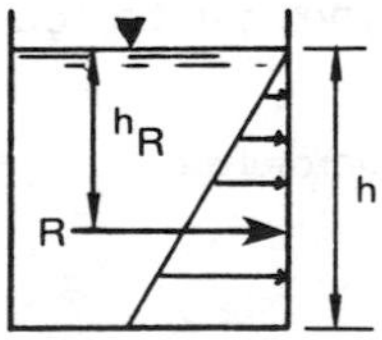

$$p = \gamma h$$

$$\bar{p} = \tfrac{1}{2}\gamma h$$

$$R = \bar{p}A$$

$$h_R = \left(\tfrac{2}{3}\right)h$$

PRESSURE ON SUBMERGED PLANE SURFACES

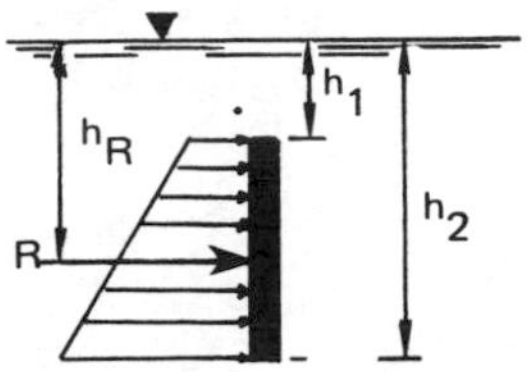

h_c = distance from surface to centroid of plane, as measured parallel to plane's surface

$$\bar{p} = \gamma h_c$$

$$R = \bar{p}A$$

$h_R = h_c + \left(\frac{I}{Ah_c}\right)$ as measured parallel to plane's surface

MANOMETERS

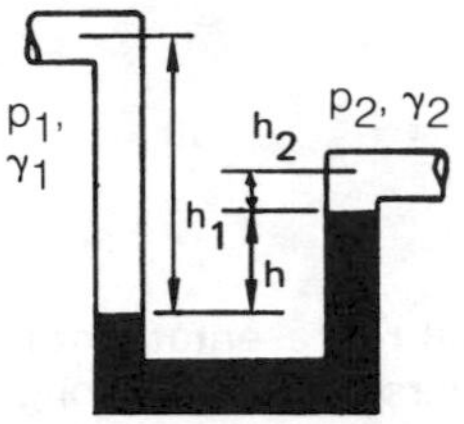

$$\Delta p = p_1 - p_2$$

$$\Delta p = \gamma_m h - \gamma_1 h_1 + \gamma_2 h_2$$

BUOYANCY (ARCHIMEDES' PRINCIPLE)

The buoyant force is equal to the weight of the displaced fluid.

SPEED OF SOUND IN A FLUID

$$c = \sqrt{kg_c RT} \quad \text{(gases)}$$

$$c = \sqrt{\frac{Eg_c}{\rho}} \quad \text{(liquids)}$$

E (bulk modulus) $\approx 4.3 \times 10^7$ lbf/ft^2 for water

CONTINUITY EQUATION

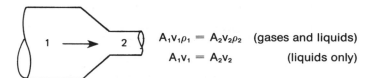

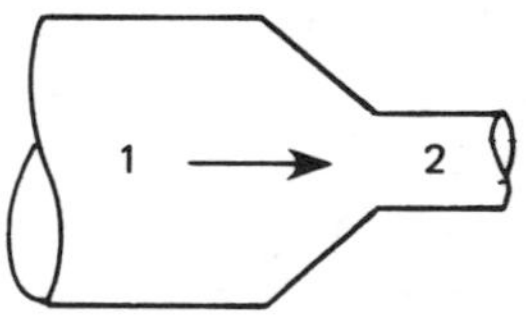

$$A_1 v_1 \rho_1 = A_2 v_2 \rho_2 \quad \text{(gases and liquids)}$$

$$A_1 v_1 = A_2 v_2 \quad \text{(liquids only)}$$

HYDRAULIC RADIUS

The hydraulic radius is defined as the area in flow divided by the wetted perimeter. The wetted perimeter does not include free fluid surface.

$$r_h = \frac{\text{area in flow}}{\text{wetted perimeter}} = D_e/4$$

BERNOULLI EQUATION

$$\frac{p}{\rho} + \frac{v^2}{2g_c} + z\left(\frac{g}{g_c}\right) = \text{total head}$$

$$\frac{p}{\gamma} = \text{pressure (static) head}$$

$$\frac{v^2}{2g} = \text{velocity (dynamic) head}$$

$$z = \text{potential (gravitational) head}$$

CONSERVATION OF ENERGY

$$\frac{p_1}{\rho} + \frac{v_1^2}{2g_c} + z_1\left(\frac{g}{g_c}\right) + w_{pump} =$$

$$\frac{p_2}{\rho} + \frac{v_2^2}{2g_c} + z_2\left(\frac{g}{g_c}\right) + h_f\left(\frac{g}{g_c}\right) + w_{turbine}$$

w_{pump} and $w_{turbine}$ are on a per-pound basis with units of (ft-lbf/lbm).

FLUID STATICS AND DYNAMICS

REYNOLDS NUMBER

$$N_{Re} = \frac{vD\rho}{\mu g_c} = \frac{vD}{\nu}$$

DARCY FRICTION LOSS AND MINOR LOSS

$$h_f = \frac{fLv^2}{2Dg} = \frac{Kv^2}{2g} \quad \text{[in feet]}$$

f is the Darcy friction factor.
K is the friction loss coefficient.

HYDRAULIC GRADE LINE

The hydraulic grade line is a graphical representation of the sum of the static and potential heads versus position along the pipeline.

$$\text{hydraulic grade} = z + \frac{p}{\gamma}$$

TORRICELLI'S EQUATION – DISCHARGE

$$C_d = C_v C_c$$

$$v_o = C_v\sqrt{2gh}$$

$$y = \tfrac{1}{2}gt^2$$

$$x = v_o t$$

$$\dot{V} = C_d A_o v_o$$

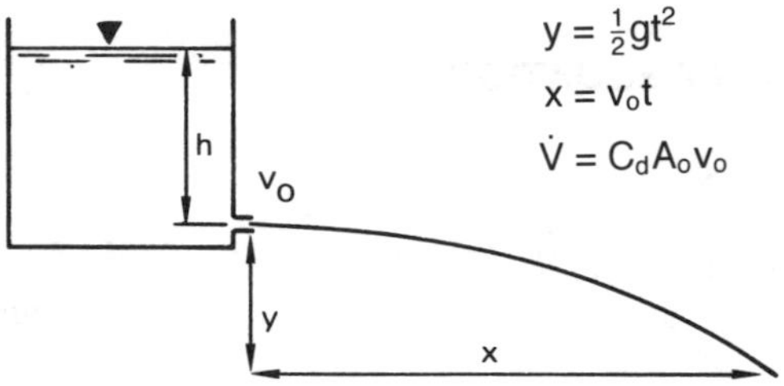

For a tank with a constant cross-sectional area, the time required to lower the fluid level from level h_1 to h_2 is

$$t = \frac{2A_t(\sqrt{h_1} - \sqrt{h_2})}{C_d A_o\sqrt{2g}}$$

VENTURI METER

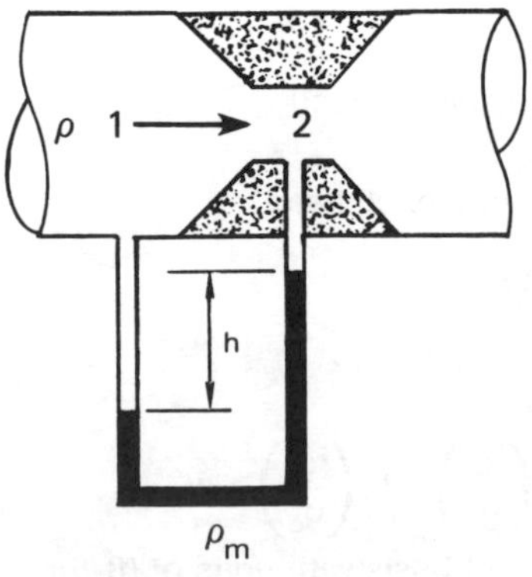

$$\beta = \frac{D_2}{D_1} = \sqrt{\frac{A_2}{A_1}}$$

$$F_{va} = \frac{1}{\sqrt{1 - (\beta)^4}}$$

$$v_2 = F_{va}\sqrt{\frac{2gh(\rho_m - \rho)}{\rho}}$$

$$C_d = C_c C_v$$

$C_c \approx 1$ for venturi meters

$$C_f = C_d F_{va}$$

$$Q = C_f A_2\sqrt{\frac{2gh(\rho_m - \rho)}{\rho}}$$

$\rho \approx 0$ for air

ORIFICE PLATE

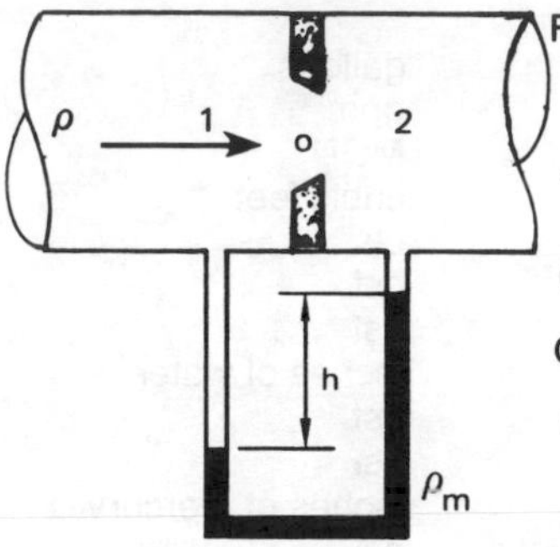

$$F_{va} = \frac{1}{\sqrt{1 - C_c^2\left(\frac{A_o}{A_1}\right)^2}}$$

$$v_2 = F_{va}C_v\sqrt{\frac{2gh(\rho_m - \rho)}{\rho}}$$

$$C_f = F_{va}C_d$$

$$C_d = C_c C_v$$

$$Q = C_c A_o v_2 = A_o C_f\sqrt{\frac{2gh(\rho_m - \rho)}{\rho}}$$

Q is in cfs

IMPULSE-MOMENTUM

Q is in cfs

$$\dot{m} = \frac{Q\rho}{g_c}$$

$$R_x = p_2A_2\cos\phi - p_1A_1 + \dot{m}(v_2\cos\phi - v_1)$$

$$R_y = A_2p_2\sin\phi + \dot{m}v_2\sin\phi$$

p's are gage pressures

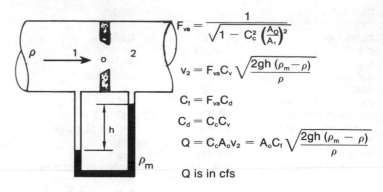

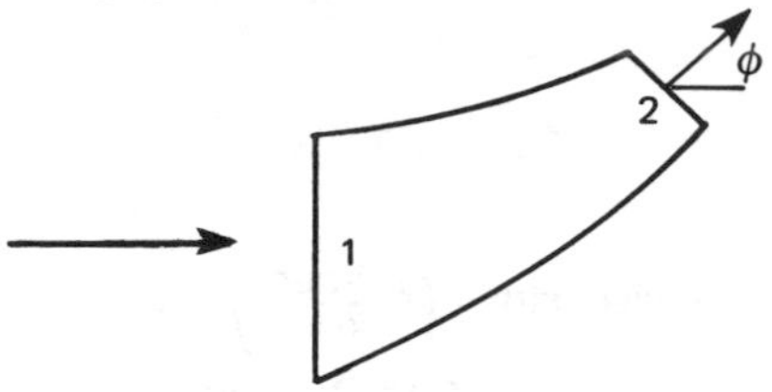

LIFT AND DRAG

$$L = \frac{C_L v^2\rho A}{2g_c}$$

$$D = \frac{C_D v^2\rho A}{2g_c}$$

A is the projected area (circle for a sphere).

SIMILARITY

For fans, pumps, turbines, drainage through holes in tanks, closed-pipe flow with no free surfaces (in the turbulent region with the same relative roughness), and for completely submerged objects such as torpedoes, airfoils, and submarines, performance of models and prototypes can be correlated by equating Reynolds numbers.

$$(N_{Re})_{model} = (N_{Re})_{prototype}$$

NET POSITIVE SUCTION HEAD

$$\text{NPSHA} = h_{\text{atmos}} + h_{\text{static}} - h_{\text{friction}} - h_{\text{vapor}}$$
$$= h_{\text{pressure (i)}} + h_{\text{velocity (i)}} - h_{\text{vapor}}$$
$$\frac{\text{NPSHR}_2}{\text{NPSHR}_1} = \left(\frac{Q_2}{Q_1}\right)^2$$

PUMPING POWER (HEAD ADDED)

$$E_A = h_A\left(\frac{g}{g_c}\right)$$
$$= \frac{p_d}{\rho} - \frac{p_i}{\rho} + \frac{v_d^2}{2g_c} - \frac{v_i^2}{2g_c} + z_d\left(\frac{g}{g_c}\right) - z_i\left(\frac{g}{g_c}\right)$$

HYDRAULIC HORSEPOWER EQUATIONS[a]

	Q in gpm	$\dot{m}$ lbm/sec	$\dot{V}$ in cfs
h_A in feet	$\frac{h_A Q(SG)}{3956}$	$\frac{h_A \dot{m}}{550} \times \frac{g}{g_c}$	$\frac{h_A \dot{V}(SG)}{8.814}$
Δp in psi	$\frac{\Delta p Q}{1714}$	$\frac{\Delta p \dot{m}}{(238.3)(SG)} \times \frac{g}{g_c}$	$\frac{\Delta p \dot{V}}{3.819}$
Δp in psf	$\frac{\Delta p Q}{2.468 \times 10^5}$	$\frac{\Delta p \dot{m}}{(34,320)(SG)} \times \frac{g}{g_c}$	$\frac{\Delta p \dot{V}}{550}$
W in $\frac{\text{ft-lbf}}{\text{lbm}}$	$\frac{WQ(SG)}{3956}$	$\frac{W\dot{m}}{550}$	$\frac{W\dot{V}(SG)}{8.814}$

[a]Based on $\rho_{\text{water}} = 62.4$ lbm/ft^3 and $g = 32.2$ ft/sec^2. (Multiply horsepower by 0.7457 to obtain kilowatts.)

HYDRAULIC KILOWATT EQUATIONS[a]

	Q in ℓ/s	$\dot{m}$ kg/s	$\dot{V}$ in m^3/s
h_A in meters	$\frac{(9.81)h_A Q(SG)}{1000}$	$\frac{(9.81)h_A \dot{m}}{1000}$	$(9.81)h_A\dot{V}(SG)$
Δp in kPa	$\frac{\Delta p Q}{1000}$	$\frac{\Delta p \dot{m}}{1000(SG)}$	$\Delta p \dot{V}$
W in $\frac{J}{kg}$	$\frac{WQ(SG)}{1000}$	$\frac{W\dot{m}}{1000}$	$W\dot{V}(SG)$

[a]Based on $\rho_{\text{water}} = 1000$ kg/m^3 and $g = 9.81$ m/s^2. (Multiply kilowatts by 1.341 to obtain horsepower.)

SPECIFIC SPEED

$$n_s = \frac{n\sqrt{Q}}{(h_A)^{0.75}} \quad \text{(pumps)}$$
$$n_s = \frac{n\sqrt{bhp}}{(H)^{1.25}} \quad \text{(turbines)}$$

CENTRIFUGAL PUMP IMPELLER TYPES

Approximate Range of Specific Speed (rpm)	Impeller Type
500–1000	radial vane
2000–3000	Francis (mixed) vane
4000–7000	mixed flow
9000 and above	axial flow

PUMP AFFINITY LAWS

$$\frac{Q_2}{Q_1} = \frac{n_2}{n_1}$$
$$\frac{h_2}{h_1} = \left(\frac{n_2}{n_1}\right)^2 = \left(\frac{Q_2}{Q_1}\right)^2$$
$$\frac{bhp_2}{bhp_1} = \left(\frac{n_2}{n_1}\right)^3 = \left(\frac{Q_2}{Q_1}\right)^3$$
$$\frac{Q_2}{Q_1} = \frac{d_2}{d_1}$$
$$\frac{h_2}{h_1} = \left(\frac{d_2}{d_1}\right)^2$$
$$\frac{bhp_2}{bhp_1} = \left(\frac{d_2}{d_1}\right)^3$$

PUMP SIMILARITY LAWS

$$\frac{n_1 d_1}{\sqrt{h_1}} = \frac{n_2 d_2}{\sqrt{h_2}}$$
$$\frac{Q_1}{d_1^2\sqrt{h_1}} = \frac{Q_2}{d_2^2\sqrt{h_2}}$$
$$\frac{bhp_1}{\rho_1 d_1^2 h_1^{1.5}} = \frac{bhp_2}{\rho_2 d_2^2 h_2^{1.5}}$$
$$\frac{Q_1}{n_1 d_1^3} = \frac{Q_2}{n_2 d_2^3}$$
$$\frac{bhp_1}{\rho_1 n_1^3 d_1^5} = \frac{bhp_2}{\rho_2 n_2^3 d_2^5}$$
$$\frac{n_1\sqrt{Q_1}}{(h_1)^{0.75}} = \frac{n_2\sqrt{Q_2}}{(h_2)^{0.75}}$$

FANS AND DUCTWORK

EQUATIONS AND CONVERSIONS

$$\text{inches w.g.} = \frac{\text{psig}}{0.0361} \quad \text{(static)}$$

$$= \left(\frac{V_{fpm}}{4005}\right)^2 \quad \text{(dynamic)}$$

AIR HORSEPOWER

$$p_t = p_s + p_v \quad \text{(inches w.g.)}$$

$$AHP = \frac{(\text{cfm})\, p_t}{6356}$$

SYSTEM LOSS CURVE

$$\frac{\Delta p_2}{\Delta p_1} = \left(\frac{Q_2}{Q_1}\right)^2$$

FAN LAWS

$$\frac{Q_A}{Q_B} = \left(\frac{D_A}{D_B}\right)^3 \left(\frac{n_A}{n_B}\right) = \left(\frac{D_A}{D_B}\right)^2 \sqrt{\frac{p_A}{p_B}} \sqrt{\frac{\rho_B}{\rho_A}}$$

$$\frac{p_A}{p_B} = \left(\frac{D_A}{D_B}\right)^2 \left(\frac{n_A}{n_B}\right)^2 \left(\frac{\rho_A}{\rho_B}\right)$$

$$\frac{AHP_A}{AHP_B} = \left(\frac{D_A}{D_B}\right)^5 \left(\frac{n_A}{n_B}\right)^3 \left(\frac{\rho_A}{\rho_B}\right)$$

$$= \left(\frac{D_A}{D_B}\right)^2 \left(\frac{p_A}{p_B}\right)^{1.5} \sqrt{\frac{\rho_B}{\rho_A}}$$

$$= \left(\frac{Q_A}{Q_B}\right) \left(\frac{p_A}{p_B}\right)$$

$$\frac{n_A}{n_B} = \left(\frac{D_B}{D_A}\right) \sqrt{\frac{p_A}{p_B}} \sqrt{\frac{\rho_B}{\rho_A}}$$

$$= \sqrt{\frac{Q_B}{Q_A}} \left(\frac{p_A}{p_B}\right)^{0.75} \left(\frac{\rho_B}{\rho_A}\right)^{0.75}$$

$$\frac{D_A}{D_B} = \sqrt{\frac{Q_A}{Q_B}} \left(\frac{p_B}{p_A}\right)^{0.25} \left(\frac{\rho_A}{\rho_B}\right)^{0.25}$$

EQUIVALENT DIAMETER OF RECTANGULAR DUCT

$$D_e = 1.3 \frac{(ab)^{0.625}}{(a+b)^{0.25}}$$

MINOR DUCT LOSSES

$$\Delta p = c p_v = c \left(\frac{v}{4005}\right)^2$$

WINDSTREAM POWER CONTENT

$$P_{ideal} = \frac{\pi r_{rotor}^2 \rho v^3}{2g_c} \quad \left(\frac{\text{ft-lbf}}{\text{sec}}\right)$$

$$P_{actual} = C_P P_{ideal}$$

C_P = power coefficient

FANS AND DUCTWORK

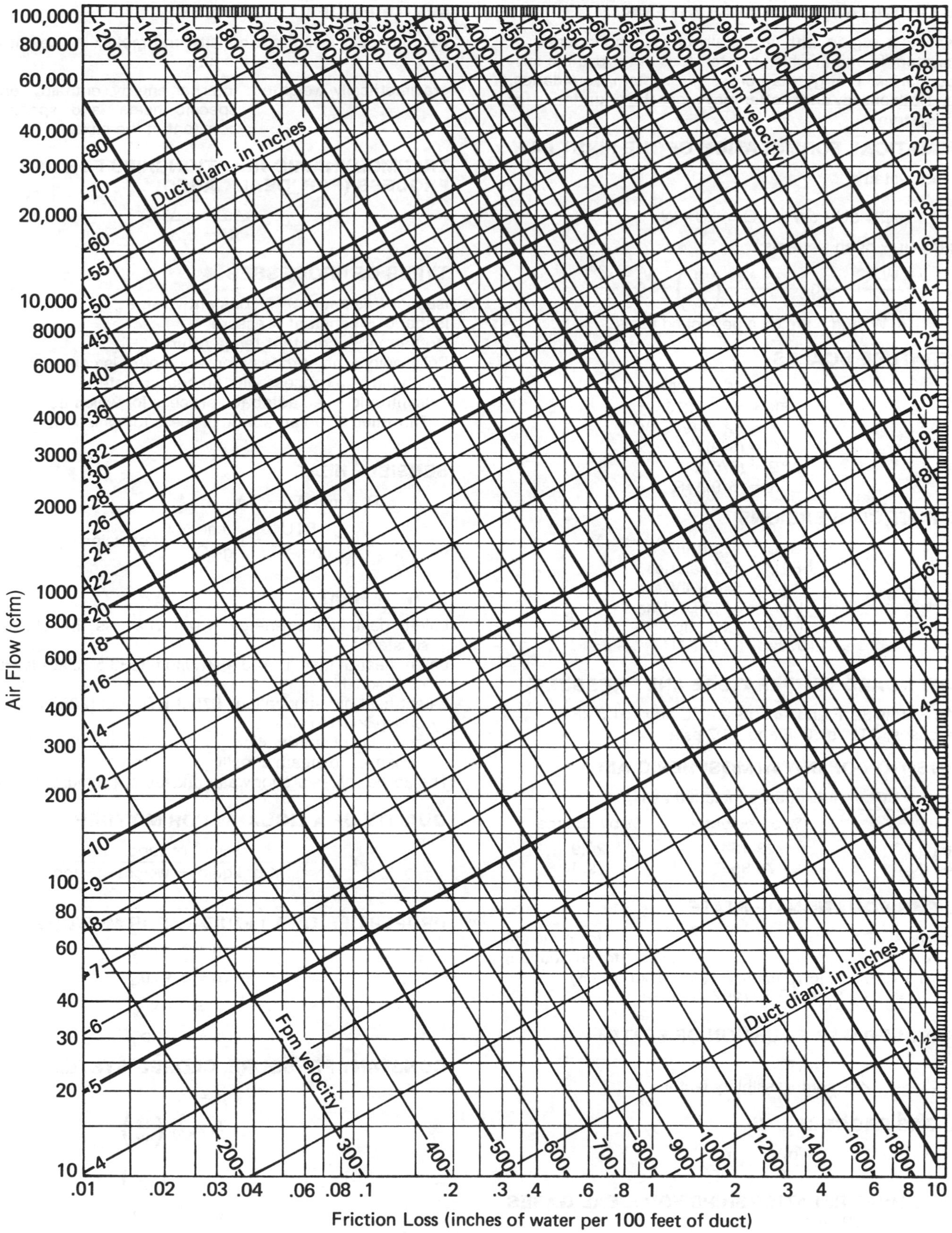

THERMODYNAMICS

TYPES OF THERMODYNAMIC SYSTEMS

A system is a region with artificially-chosen boundaries. A closed system is one in which matter does not cross the system boundaries. Energy may cross the system boundaries, however. Both energy and matter may cross the boundaries for an open system.

MATHEMATICAL FORMULATION OF THE FIRST LAW

For closed systems:

$$\delta Q = dU + \delta W \text{ or } Q = \Delta U + W$$

For steady-flow open systems:

$$q = \frac{W}{J} + h_2 - h_1 + \frac{z_2 - z_1}{J}\left(\frac{g}{g_c}\right) + \frac{v_2^2 - v_1^2}{2g_cJ}$$

(q, W, and h are on a per-pound basis.)

TEMPERATURE SCALES

$$°R = °F + 460$$
$$°K = °C + 273$$
$$°F = 32 + (9/5)\,°C$$
$$°C = (5/9)(°F - 32)$$
$$\Delta°R = (9/5)\Delta°K$$
$$\Delta°K = (5/9)\Delta°R$$

IMPORTANT CONVERSIONS

1 BTU = 778 ft-lbf
1 BTU = 252 calories
1 hp = 550 ft-lbf/sec
1 kW = 3413 BTU/hr
1 kW = 1.341 hp
1 kW-hr = 3413 BTU

STP (STANDARD TEMPERATURE AND PRESSURE)

Scientific STP is 32°F and 14.7 psia (0°C and 101.3 kPa).
For fuel gases, STP is 60°F and 14.7 psia.

COMPOSITION OF DRY ATMOSPHERIC AIR

(Rare inert gases included as N_2)

	% by weight	% by volume
Oxygen (O_2)	23.15	20.9
Nitrogen (N_2)	76.85	79.1

IDEAL GAS EQUATION OF STATE

$pV = nR^*T$ (for n moles)
$pV = mRT$ (for m pounds)
$R^* = 1545$ ft-lbf/pmole-°R (8314 J/kmole · K)
$R = 53.3$ ft/°R for air (287 J/kg · K)

COMPRESSED GAS EQUATION OF STATE

$pV = mZRT$
Z = compressibility factor

IDEAL GAS PROCESS LAW

$$\frac{p_1V_1}{T_1} = \frac{p_2V_2}{T_2}$$

SPECIFIC HEAT RELATIONSHIPS FOR IDEAL GASES

$c_p - c_v = (R/J)$
$c_p = Rk/J(k-1)$
$c_p = 0.24$ BTU/lbm-°F (1000 J/kg · K) (for air)
$c_v = 0.1714$ BTU/lbm-°F (718 J/kg · K) (for air)
$k = c_p/c_v = 1.4$ (for air)

MIXTURES OF IDEAL GASES

volumetrically weighted: density and molecular weight of the mixture

gravimetrically weighted: internal energy, enthalpy, entropy, specific heats, and specific gas constant of the mixture

THERMODYNAMIC RELATIONSHIPS FOR ANY PROCESS (FOR IDEAL GASES)

$$\Delta u = c_v\Delta T$$
$$\Delta h = c_p\Delta T$$

TYPES OF PROCESSES

Isochoric (isometric) — constant volume
Isobaric — constant pressure
Isothermal — constant temperature
Polytropic — any process for which $p(V)^n$ is constant
Adiabatic — a process with no heat transfer
- Isentropic — an adiabatic process for which $\Delta s = 0$
- Throttling — an adiabatic process for which $\Delta h = 0$

SENSIBLE HEAT

$$Q = mc(T_2 - T_1)$$

$$c = 1\,\frac{\text{BTU}}{\text{lbm-°F}} = 1\,\frac{\text{cal}}{\text{g-°C}} \quad \text{(for water)}$$

LATENT HEAT

Latent heat is the energy which changes the phase of a substance.
Latent heat of fusion: 143.4 BTU/lbm (333.5 kJ/kg) for turning ice to liquid
Latent heat of vaporization: 970.3 BTU/lbm (2256.7 kJ/kg) for turning liquid to steam
Latent heat of sublimation: 1220 BTU/lbm (2838 kJ/kg) for turning ice to steam directly
The above values are for water phases at 1 atmosphere.

QUALITY OF A LIQUID-VAPOR MIXTURE

$$x = \frac{\text{weight of vapor}}{\text{weight of liquid} + \text{weight of vapor}}$$

PROPERTIES OF A LIQUID-VAPOR MIXTURE

$$h = h_f + xh_{fg}$$
$$u = u_f + xu_{fg}$$
$$\nu = \nu_f + x\nu_{fg}$$
$$s = s_f + xs_{fg}$$

CONSTANT PRESSURE, CLOSED SYSTEMS

$$p_2 = p_1$$
$$T_2 = T_1\left(\frac{\upsilon_2}{\upsilon_1}\right)$$
$$\upsilon_2 = \upsilon_1\left(\frac{T_2}{T_1}\right)$$
$$Q = h_2 - h_1$$
$$= c_p(T_2 - T_1)$$
$$= c_v(T_2 - T_1) + p(\upsilon_2 - \upsilon_1)$$

$$u_2 - u_1 = c_v(T_2 - T_1)$$
$$= \frac{c_v p(v_2 - v_1)}{R}$$
$$= \frac{p(v_2 - v_1)}{k - 1}$$
$$W = p(v_2 - v_1)$$
$$= R(T_2 - T_1)$$
$$s_2 - s_1 = c_p ln\left(\frac{T_2}{T_1}\right)$$
$$= c_p ln\left(\frac{v_2}{v_1}\right)$$
$$h_2 - h_1 = Q$$
$$= \frac{kp(v_2 - v_1)}{k - 1}$$

CONSTANT VOLUME, CLOSED SYSTEMS

$$p_2 = p_1\left(\frac{T_2}{T_1}\right)$$
$$T_2 = T_1\left(\frac{p_2}{p_1}\right)$$
$$v_2 = v_1$$
$$Q = u_2 - u_1$$
$$= c_v(T_2 - T_1)$$
$$u_2 - u_1 = Q$$
$$= \frac{c_v v(p_2 - p_1)}{R}$$
$$= \frac{v(p_2 - p_1)}{k - 1}$$
$$W = 0$$
$$s_2 - s_1 = c_v ln\left(\frac{T_2}{T_1}\right)$$
$$= c_v ln\left(\frac{p_2}{p_1}\right)$$
$$h_2 - h_1 = c_p(T_2 - T_1)$$
$$= \frac{kv(p_2 - p_1)}{k - 1}$$

CONSTANT TEMPERATURE, CLOSED SYSTEMS

$$p_2 = p_1\left(\frac{v_1}{v_2}\right)$$
$$T_2 = T_1$$
$$v_2 = v_1\left(\frac{p_1}{p_2}\right)$$
$$Q = W$$
$$= T(s_2 - s_1)$$
$$= p_1 v_1 \, ln\left(\frac{v_2}{v_1}\right)$$
$$= RT \, ln\left(\frac{v_2}{v_1}\right)$$
$$u_2 - u_1 = 0$$
$$W = Q$$
$$= RT \, ln\left(\frac{p_1}{p_2}\right)$$
$$s_2 - s_1 = \frac{Q}{T}$$
$$= R \, ln\left(\frac{p_1}{p_2}\right)$$
$$h_2 - h_1 = 0$$

ISENTROPIC, STEADY FLOW SYSTEMS

p_2, v_2, and T_2 are the same as for isentropic, closed systems.

$$Q = 0$$
$$W = h_1 - h_2$$
$$= c_p T_1\left[1 - \left(\frac{p_2}{p_1}\right)^{\frac{k-1}{k}}\right]$$
$$u_2 - u_1 = c_v(T_2 - T_1)$$
$$h_1 - h_2 = W$$
$$= c_p(T_2 - T_1)$$
$$= \frac{k(p_2 v_2 - p_1 v_1)}{k - 1}$$
$$s_2 - s_1 = 0$$

POLYTROPIC, CLOSED SYSTEMS

To use with isentropic, closed systems, substitute k for n in all equations.

$$p_2 = p_1\left(\frac{v_1}{v_2}\right)^n$$
$$= p_1\left(\frac{T_2}{T_1}\right)^{\frac{n}{n-1}}$$
$$T_2 = T_1\left(\frac{v_1}{v_2}\right)^{n-1}$$
$$= T_1\left(\frac{p_2}{p_1}\right)^{\frac{n-1}{n}}$$
$$v_2 = v_1\left(\frac{p_1}{p_2}\right)^{\frac{1}{n}}$$
$$= v_1\left(\frac{T_1}{T_2}\right)^{\frac{1}{n-1}}$$
$$Q = \frac{c_v(n - k)(T_2 - T_1)}{n - 1}$$
$$u_2 - u_1 = c_v(T_2 - T_1)$$
$$= \frac{p_2 v_2 - p_1 v_1}{n - 1}$$
$$W = \frac{R(T_1 - T_2)}{n - 1}$$
$$= \frac{p_1 v_1 - p_2 v_2}{n - 1}$$
$$= \frac{p_1 v_1}{n - 1}\left[1 - \left(\frac{p_2}{p_1}\right)^{\frac{n-1}{n}}\right]$$
$$s_2 - s_1 = \frac{c_v(n - k)}{n - 1}\left[ln\left(\frac{T_2}{T_1}\right)\right]$$
$$h_2 - h_1 = c_p(T_2 - T_1)$$
$$= \frac{n(p_2 v_2 - p_1 v_1)}{n - 1}$$

POLYTROPIC, STEADY-FLOW SYSTEMS

p_2, v_2, and T_2 are the same as for polytropic, closed systems.

$$q = \frac{c_v(n-k)(T_2 - T_1)}{n-1}$$

$$W = h_1 - h_2$$

$$= \frac{nc_v(1-k)T_1}{n-1}\left[1 - \left(\frac{p_2}{p_1}\right)^{\frac{n-1}{n}}\right]$$

$$u_2 - u_1 = c_v(T_2 - T_1)$$

$$h_2 - h_1 = -W$$

$$= c_p(T_2 - T_1)$$

$$= \frac{n(p_2v_2 - p_1v_1)}{n-1}$$

$$s_2 - s_1 = \frac{c_v(n-k)}{n-1}\left[\ln\left(\frac{T_2}{T_1}\right)\right]$$

THROTTLING, STEADY-FLOW SYSTEMS

$$p_1v_1 = p_2v_2$$

$$p_2 < p_1$$

$$v_2 > v_1$$

$$T_2 = T_1$$

$$q = 0$$

$$W = 0$$

$$u_2 - u_1 = 0$$

$$h_2 - h_1 = 0$$

$$s_2 - s_1 = R\ln\left(\frac{p_1}{p_2}\right)$$

$$= R\ln\left(\frac{v_2}{v_1}\right)$$

ENERGY, WORK, AND POWER CONVERSIONS

multiply	by	to get
BTU	3.929×10^{-4}	hp-hrs
BTU	778.3	ft-lbf
BTU	2.930×10^{-4}	kW-hrs
BTU	1.0×10^{-5}	therms
BTU/hr	0.2161	ft-lbf/sec
BTU/hr	3.929×10^{-4}	hp
BTU/hr	0.2930	watts
ft-lbf	1.285×10^{-3}	BTU
ft-lbf	3.766×10^{-7}	kW-hrs
ft-lbf	5.051×10^{-7}	hp-hrs
ft-lbf/sec	4.6272	BTU/hr
ft-lbf/sec	1.818×10^{-3}	hp
ft-lbf/sec	1.356×10^{-3}	kW
hp	2545.0	BTU/hr
hp	550	ft-lbf/sec
hp	0.7457	kW
hp-hr	2545.0	BTU
hp-hr	1.976×10^{6}	ft-lbf
hp-hr	0.7457	kW-hrs
kW	1.341	hp
kW	3412.9	BTU/hr
kW	737.6	ft-lbf/sec
kW	3412.9	BTU

GENERAL POWER CYCLE

The general power cycle moves clockwise on the p-v and T-s diagrams.

1→2: compression
2→3: heat addition
3→4: expansion
4→1: heat rejection

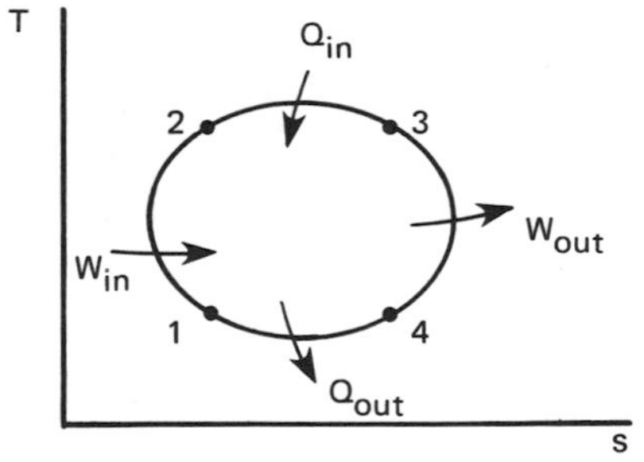

THERMAL EFFICIENCY OF THE ENTIRE CYCLE

$$\eta_{th} = \frac{Q_{in} - Q_{out}}{Q_{in}} = \frac{W_{out} - W_{in}}{Q_{in}}$$

CARNOT CYCLE EFFICIENCY

For the Carnot cycle, the thermal efficiency does not depend on the working fluid. It can be calculated directly from the two temperature extremes.

$$\eta_{th,\ Carnot} = \frac{T_{high} - T_{low}}{T_{high}}$$

ISENTROPIC EFFICIENCY OF A PUMP

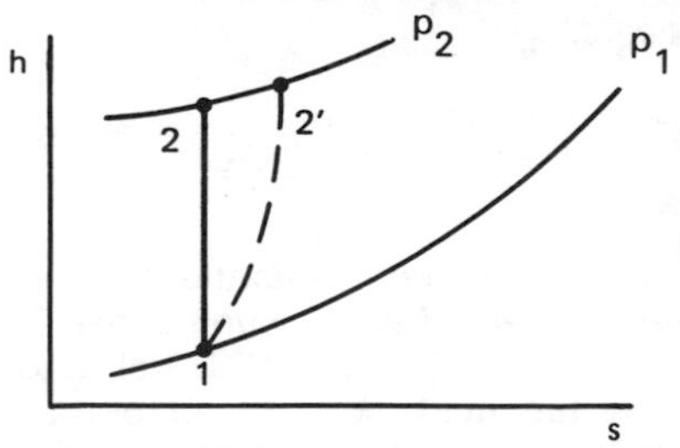

$$\eta = \frac{h_2 - h_1}{h_2' - h_1}$$

$$h_2' = h_1 + \frac{h_2 - h_1}{\eta}$$

ISENTROPIC EFFICIENCY OF A TURBINE

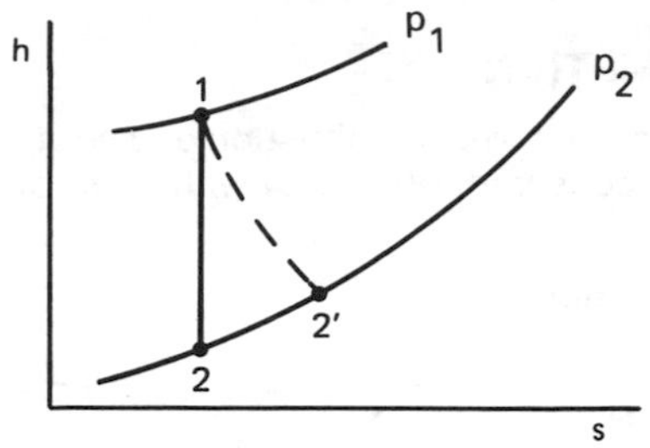

$$\eta = \frac{h_1 - h_2'}{h_1 - h_2}$$

$$h_2' = h_1 - \eta\,(h_1 - h_2)$$

RANKINE CYCLE WITH SUPERHEATING

$$W_{turb} = h_d - h_e$$

$$W_{pump} = \upsilon_f\,(p_a - p_f) = h_a - h_f$$

$$Q_{in} = h_d - h_a$$

$$Q_{out} = h_e - h_f$$

The thermal efficiency of the entire cycle is

$$\eta_{th} = \frac{Q_{in} - Q_{out}}{Q_{in}} = \frac{W_{turb} - W_{pump}}{Q_{in}}$$

$$= \frac{(h_d - h_a) - (h_e - h_f)}{h_d - h_a}$$

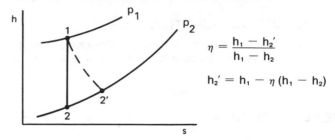

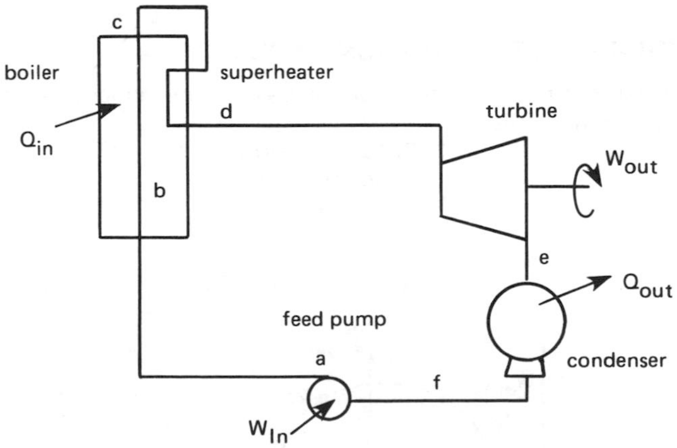

If the Rankine cycle gives efficiencies for the turbine and the pump, calculate all quantities as if those efficiencies were 100%. Then modify h_e and h_a prior to recalculating the thermal efficiency.

$$h_e' = h_d - \eta_{turb}\,(h_d - h_e)$$

$$h_a' = h_f + \frac{h_a - h_f}{\eta_{pump}}$$

$$W'_{turb} = h_d - h'_e$$

$$W'_{pump} = h'_a - h_f$$

$$Q'_{in} = h_d - h'_a$$

HEAT OF COMBUSTION

The heat of combustion of a fuel (also known as the 'heating value') is the amount of energy given off when a unit of fuel is burned. Units of heating value are BTU/lbm, BTU/gal, and BTU/ft^3 depending on whether the fuel is solid, liquid, or gas. The higher heating value (HHV) includes the heat of vaporization of the water vapor formed.

RATE OF FUEL CONSUMPTION

The rate of fuel consumption in internal combustion engines is known as the 'specific fuel consumption' (SFC) with units of lbm/hp-hr.

$$\text{hourly fuel consumption} = (\text{hp})(\text{SFC})$$

THE PLAN FORMULA

$$\text{hp} = \frac{\text{pLAN}}{33{,}000}$$

$$\text{N} = \frac{(2\text{n})(\text{no. cylinders})}{\text{no. strokes per cycle}}$$

p is the mean effective pressure in psig
L is the stroke length in feet
A is the bore area in in^2
N is the number of power strokes per minute
n is the engine speed in rpm

HORSEPOWER VERSUS TORQUE

$$(\text{hp})(5252) = (T_{\text{ft-lbf}})(\text{rpm})$$

FLOW THROUGH NOZZLES

$$v = \sqrt{2g_c J(h_1 - h_2)}$$

GENERAL REFRIGERATION CYCLE

The general refrigeration cycle moves counter-clockwise on the p-v and T-s diagrams

1→2: compression
2→3: heat rejection
3→4: expansion
4→1: heat addition

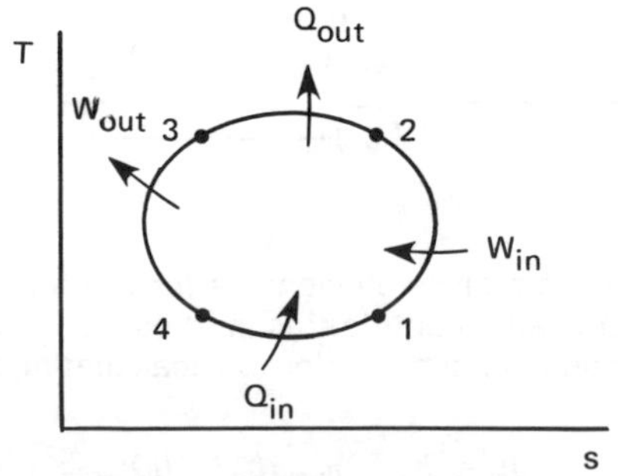

REFRIGERATION UNITS

1 ton = 200 BTU/min; 12,000 BTU/hr; 3517 W

HEAT PUMPS

A heat pump operates on a refrigeration cycle using refrigeration equipment. The only difference is the use to which a heat pump is put. The purpose of a heat pump is to warm the region in which the heat rejection coils are located.

COEFFICIENT OF PERFORMANCE

Efficiencies are not calculated for refrigeration cycles. Rather, coefficients of performance (COP) are used to compare different cycles.

$$\text{COP}_{\text{refrigerator}} = \frac{Q_{\text{absorbed}}}{W_{\text{compression}}}$$

$$= \frac{Q_{\text{absorbed}}}{Q_{\text{rejected}} - Q_{\text{absorbed}}}$$

$$\text{COP}_{\text{heat pump}} = \frac{Q_{\text{rejected}}}{W_{\text{compression}}}$$

$$= \frac{Q_{\text{absorbed}} + W_{\text{compression}}}{W_{\text{compression}}}$$

$$= \text{COP}_{\text{refrigerator}} + 1$$

CARNOT CYCLE COP

$$\text{COP}_{\text{refrigerator}} = \frac{T_{\text{low}}}{T_{\text{high}} - T_{\text{low}}}$$

$$\text{COP}_{\text{heat pump}} = \frac{T_{\text{high}}}{T_{\text{high}} - T_{\text{low}}}$$

$$= \text{COP}_{\text{refrigerator}} + 1$$

COMPRESSIBLE FLUID DYNAMICS

GENERAL FLOW EQUATION

$$\frac{v_1^2}{2g_c} + Jh_1 = \frac{v_2^2}{2g_c} + Jh_2 \qquad (\Delta z \approx 0,\ h \text{ is enthalpy})$$

SPEED OF SOUND

$$c = \sqrt{kg_cRT}$$

MACH NUMBER

$$M = \frac{v}{c}$$

ISENTROPIC FLOW PARAMETERS

$$\left(\frac{T_T}{T_2}\right) = \tfrac{1}{2}(k-1)M_2^2 + 1$$

$$\left(\frac{p_T}{p}\right) = \left[\tfrac{1}{2}(k-1)M^2 + 1\right]^{\frac{k}{(k-1)}}$$

$$\left(\frac{\rho_T}{\rho}\right) = \left[\tfrac{1}{2}(k-1)M^2 + 1\right]^{\frac{1}{(k-1)}}$$

$$\left(\frac{A}{A^*}\right) = \frac{1}{M}\left[\frac{\tfrac{1}{2}(k-1)M^2 + 1}{\tfrac{1}{2}(k-1) + 1}\right]^{\frac{k+1}{2(k-1)}}$$

$$\left(\frac{v}{c^*}\right) = \sqrt{\frac{\tfrac{1}{2}(k+1)M^2}{\tfrac{1}{2}(k-1)M^2 + 1}}$$

CRITICAL PRESSURE RATIO

$$R_{cp} = \left(\frac{2}{k+1}\right)^{\frac{k}{k-1}}$$

ROCKET PERFORMANCE

$$\dot{m} = \rho_e v_e A_e$$

thrust

$$F = \frac{\dot{m}\,v_e}{g_c} + A_e(p_e - p_a) = \frac{\dot{m}\,v_{eff}}{g_c}$$

characteristic velocity

$$v_{char} = \frac{g_c p_T A_t}{\dot{m}} = \frac{v_{eff}}{C_f}$$

coefficient of thrust

$$C_f = \frac{F}{p_T A_t}$$

total impulse

$$I = Ft$$

STEAM FLOW (NOZZLE FLOW)

$$v_2 = \sqrt{2g_cJ(h_1 - h_2) + v_1^2}$$

$$\eta_{nozzle} = \frac{\Delta h_{actual}}{\Delta h_{ideal}} = \left(\frac{v_{actual}}{v_{ideal}}\right)^2$$

COMPRESSIBLE FLUID DYNAMICS

ISENTROPIC FLOW FACTORS FOR k = 1.4

M = Mach number, P = pressure, TP = total pressure, D = density, TD = total density, T = temperature, TT = total temperature, A = area at a point, A* = throat area for M = 1, v = velocity, and c* = speed of sound at throat.

M	P/TP	D/TD	T/TT	A/A*	V/C*
.00	1.0000	1.0000	1.0000	-------	.0000
.01	.9999	1.0000	1.0000	57.8737	.0110
.02	.9997	.9998	.9999	28.9420	.0219
.03	.9994	.9996	.9998	19.3005	.0329
.04	.9989	.9992	.9997	14.4815	.0438
.05	.9983	.9988	.9995	11.5914	.0548
.06	.9975	.9982	.9993	9.6659	.0657
.07	.9966	.9976	.9990	8.2915	.0766
.08	.9955	.9968	.9987	7.2616	.0876
.09	.9944	.9960	.9984	6.4613	.0985
.10	.9930	.9950	.9980	5.8218	.1094
.11	.9916	.9940	.9976	5.2992	.1204
.12	.9900	.9928	.9971	4.8643	.1313
.13	.9883	.9916	.9966	4.4969	.1422
.14	.9864	.9903	.9961	4.1824	.1531
.15	.9844	.9888	.9955	3.9103	.1639
.16	.9823	.9873	.9949	3.6727	.1748
.17	.9800	.9857	.9943	3.4635	.1857
.18	.9776	.9840	.9936	3.2779	.1965
.19	.9751	.9822	.9928	3.1123	.2074
.20	.9725	.9803	.9921	2.9635	.2182
.21	.9697	.9783	.9913	2.8293	.2290
.22	.9668	.9762	.9904	2.7076	.2398
.23	.9638	.9740	.9895	2.5968	.2506
.24	.9607	.9718	.9886	2.4956	.2614
.25	.9575	.9694	.9877	2.4027	.2722
.26	.9541	.9670	.9867	2.3173	.2829
.27	.9506	.9645	.9856	2.2385	.2936
.28	.9470	.9619	.9846	2.1656	.3043
.29	.9433	.9592	.9835	2.0979	.3150
.30	.9395	.9564	.9823	2.0351	.3257
.31	.9355	.9535	.9811	1.9765	.3364
.32	.9315	.9506	.9799	1.9218	.3470
.33	.9274	.9476	.9787	1.8707	.3576
.34	.9231	.9445	.9774	1.8229	.3682
.35	.9188	.9413	.9761	1.7780	.3788
.36	.9143	.9380	.9747	1.7358	.3893
.37	.9098	.9347	.9734	1.6961	.3999
.38	.9052	.9313	.9719	1.6587	.4104
.39	.9004	.9278	.9705	1.6234	.4209
.40	.8956	.9243	.9690	1.5901	.4313
.41	.8907	.9207	.9675	1.5587	.4418
.42	.8857	.9170	.9659	1.5289	.4522
.43	.8807	.9132	.9643	1.5007	.4626
.44	.8755	.9094	.9627	1.4740	.4729
.45	.8703	.9055	.9611	1.4487	.4833
.46	.8650	.9016	.9594	1.4246	.4936
.47	.8596	.8976	.9577	1.4018	.5038
.48	.8541	.8935	.9560	1.3801	.5141
.49	.8486	.8894	.9542	1.3595	.5243
.50	.8430	.8852	.9524	1.3398	.5345
.51	.8374	.8809	.9506	1.3212	.5447
.52	.8317	.8766	.9487	1.3034	.5548
.53	.8259	.8723	.9468	1.2865	.5649
.54	.8201	.8679	.9449	1.2703	.5750
.55	.8142	.8634	.9430	1.2549	.5851
.56	.8082	.8589	.9410	1.2403	.5951
.57	.8022	.8544	.9390	1.2263	.6051
.58	.7962	.8498	.9370	1.2130	.6150
.59	.7901	.8451	.9349	1.2003	.6249
.60	.7840	.8405	.9328	1.1882	.6348
.61	.7778	.8357	.9307	1.1767	.6447
.62	.7716	.8310	.9286	1.1656	.6545
.63	.7654	.8262	.9265	1.1551	.6643
.64	.7591	.8213	.9243	1.1451	.6740
.65	.7528	.8164	.9221	1.1356	.6837
.66	.7465	.8115	.9199	1.1265	.6934
.67	.7401	.8066	.9176	1.1179	.7031
.68	.7338	.8016	.9153	1.1097	.7127
.69	.7274	.7966	.9131	1.1018	.7223
.70	.7209	.7916	.9107	1.0944	.7318
.71	.7145	.7865	.9084	1.0873	.7413
.72	.7080	.7814	.9061	1.0806	.7508
.73	.7016	.7763	.9037	1.0742	.7602
.74	.6951	.7712	.9013	1.0681	.7696
.75	.6886	.7660	.8989	1.0624	.7789
.76	.6821	.7609	.8964	1.0570	.7883
.77	.6756	.7557	.8940	1.0519	.7975
.78	.6691	.7505	.8915	1.0471	.8068
.79	.6625	.7452	.8890	1.0425	.8160
.80	.6560	.7400	.8865	1.0382	.8251
.81	.6495	.7347	.8840	1.0342	.8343
.82	.6430	.7295	.8815	1.0305	.8433
.83	.6365	.7242	.8789	1.0270	.8524
.84	.6300	.7189	.8763	1.0237	.8614
.85	.6235	.7136	.8737	1.0207	.8704
.86	.6170	.7083	.8711	1.0179	.8793
.87	.6106	.7030	.8685	1.0153	.8882
.88	.6041	.6977	.8659	1.0129	.8970
.89	.5977	.6924	.8632	1.0108	.9058
.90	.5913	.6870	.8606	1.0089	.9146
.91	.5849	.6817	.8579	1.0071	.9233
.92	.5785	.6764	.8552	1.0056	.9320
.93	.5721	.6711	.8525	1.0043	.9406
.94	.5658	.6658	.8498	1.0031	.9493
.95	.5595	.6604	.8471	1.0021	.9578
.96	.5532	.6551	.8444	1.0014	.9663
.97	.5469	.6498	.8416	1.0008	.9748
.98	.5407	.6445	.8389	1.0003	.9832
.99	.5345	.6392	.8361	1.0001	.9916
1.00	.5283	.6339	.8333	1.0000	1.0000

COMBUSTION

Quantities in ( ) are volumetric percents from Orsat analysis. Quantities in [ ] are gravimetric percents from ultimate analysis.

ELEMENTS AND COMPOUNDS

name	molecular weight
carbon	12.0
hydrogen	2.01
nitrogen	28.01
oxygen	32.0
sulfur	32.06

COMPOSITION OF AIR

	% by weight	% by volume
oxygen	23.15	20.9
nitrogen and inerts	76.85	79.1

HEAT OF COMBUSTION

(in BTU/lbm)

$$HV = 14{,}093[C] + 60{,}958\left([H_2] - \frac{[O_2]}{8}\right) + 3983[S]$$

FLUE GAS ANALYSIS

$$\frac{\text{lbm actual air}}{\text{lbm fuel}} = \frac{3.04\,(N_2)[C]}{(CO_2) + (CO)}$$

$$\%\text{ excess air} = \frac{100\,((O_2) - 0.5\,(CO))}{0.264\,(N_2) - [(O_2) - 0.5\,(CO)]}$$

$$(CO_2)_{\text{theoretical}} = \frac{100\,(CO_2)_{\text{actual}}}{1 - 4.76\,(O_2)}$$

$$\frac{\text{lbm dry flue gas}}{\text{lbm solid fuel}} = \frac{\{11(CO_2) + 8(O_2) + 7\{(CO) + (N_2)\}\}\left\{[C] + \left(\frac{[S]}{1.833}\right)\right\}}{3((CO_2) + (CO))}$$

PROPERTIES OF COMMON FUELS

	Gasoline	Octane	Propane	Ethanol	No. 1 Diesel	No. 2 Diesel
Chemical formula	-	C_8H_{18}	C_3H_8	C_2H_5OH	—	—
Molecular weight	≈126	114	44	46	≈170	≈184
Carbon % by weight	—	84	82	52	—	—
Hydrogen % by weight	—	16	18	13	—	—
Oxygen % by weight	—	—	—	35	—	—
Heating value						
Higher BTU/lbm	20,260	20,590	21,646	12,800	19,240	19,110
Lower BTU/lbm	18,900	19,100	19,916	11,500	18,250	18,000
BTU/gal (lower)	116,485	111,824	81,855	76,152	133,332	138,110
Latent heat of Vaporization BTU/lbm	142	141	147	361	115	105
Specific gravity	0.739	0.702	0.493	0.794	0.876	0.920
Research octane	85-94	100	112	106		—
Motor octane	77-86	100	97	89	10-30	
Cetane number	10 to 20	—	—	−20 to 8	≈45	—
Stoichiometric Mass A/F ratio	14.7	15.1	—	9.0	—	—
Distillation Temperature (°F)	90-410	—	—	173	340-560	—
Flammability Limits (volume percent)	1.4 to 7.6	—	—	4.3 to 19	—	—

HEAT TRANSFER

THERMAL CONDUCTIVITY

$$k_T = k_o(1 + \gamma T)$$

CONDUCTION THROUGH SLABS

$$q = \frac{kA\Delta T}{L}$$

CONDUCTION THROUGH SANDWICHES

$$q = \frac{A\Delta T}{\sum\left(\frac{L_i}{k_i}\right)} \quad \text{(no films)}$$

$$q = \frac{A\Delta T}{\sum\left(\frac{L_i}{k_i}\right) + \sum\left(\frac{1}{h_j}\right)} \quad \text{(films)}$$

COEFFICIENT OF HEAT TRANSFER (CONDUCTIVITY)

$$U = \frac{q}{A\Delta T} \quad \text{(general rule)}$$

$$U = \frac{1}{\sum\left(\frac{L_i}{k_i}\right) + \sum\left(\frac{1}{h_j}\right)} \quad \text{(sandwiches with films)}$$

LOGARITHMIC MEAN AREA

$$A_m = \frac{A_o - A_i}{ln\left(\frac{A_o}{A_i}\right)}$$

HOLLOW CYLINDER

$$q = \frac{2\pi kL\Delta T}{ln\left(\frac{r_o}{r_i}\right)}$$

$$A_m = \frac{2\pi L(r_o - r_i)}{ln\left(\frac{r_o}{r_i}\right)}$$

INSULATED PIPE

$$q = \frac{2\pi L\Delta T}{\frac{1}{r_a h_a} + \frac{ln\left(\frac{r_b}{r_a}\right)}{k_{pipe}} + \frac{1}{r_b h_b} + \frac{ln\left(\frac{r_c}{r_b}\right)}{k_{in}} + \frac{1}{r_c h_c}}$$

Note: h_b may not be present.

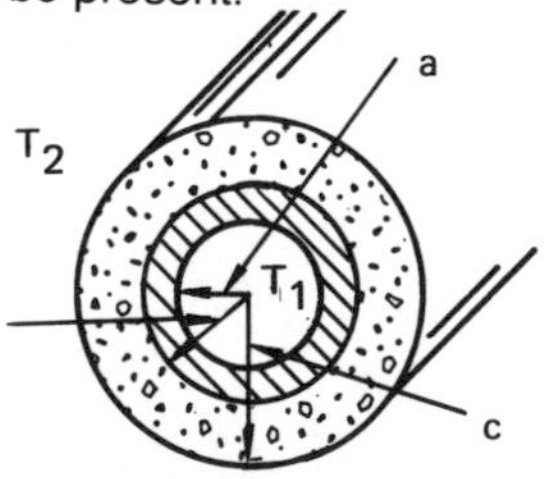

CRITICAL THICKNESS OF INSULATION

$$r_{critical} = \frac{k_{insulation}}{h}$$

NEWTON'S LAW OF COOLING

$$T_t = T_\infty + (T_o - T_\infty)e^{-rt}$$

$$t = \left(\frac{1}{r}\right) ln\left(\frac{T_t - T_\infty}{T_o - T_\infty}\right)$$

The rate constant, r, must be determined from heat loss data.

BIOT NUMBER

$$N_{Bi} = \frac{hL}{k}$$

THERMAL DIFFUSIVITY

$$a = \frac{k}{\rho C_p}$$

FOURIER NUMBER

$$N_{Fo} = \frac{kt}{\rho C_p L^2} = \frac{at}{L^2}$$

TRANSIENT HEAT FLOW

(only for $N_{Bi} < 10$)

$$T_t = T_\infty + \Delta T e^{-N_{Bi} N_{Fo}}$$

$$q_t = hA_s \Delta T e^{-N_{Bi} N_{Fo}}$$

$$\Delta T = T_{initial} - T_{environment}$$

REYNOLDS' NUMBER

$$N_{Re} = \frac{Dv\rho}{g_c\mu} = \frac{Dv}{\nu} \quad \text{[use consistent units]}$$

PRANDTL NUMBER

$$N_{Pr} = \frac{C_p\mu}{k} \quad \text{[use consistent units]}$$

GRASHOFF NUMBER

$$N_{Gr} = \frac{L^3\rho^2\beta\Delta Tg}{\mu^2} \quad \text{[use consistent units]}$$

$$\Delta T = T_{surface} - T_{environment}$$

Evaluate film properties at $\frac{1}{2}(T_{surface} + T_{environment})$.

NUSSELT EQUATION FOR FORCED CONVECTION IN PIPES

$$\frac{hD}{k} = 0.0225\,(N_{Re})^{0.8}(N_{Pr})^n$$

$n = 0.3$ for heat flow out of pipe

$= 0.4$ for heat flow into pipe

NATURAL CONVECTION

$$\frac{hL}{k} = C(N_{Gr}N_{Pr})^n \text{ for } N_{Pr} > 0.6$$

C and n depend on the configuration.

Film properties are evaluated at ½ $(T_{surface} + T_{environment})$

LOGARITHMIC MEAN TEMPERATURE DIFFERENCE

$$\Delta T_m = \frac{\Delta T_A - \Delta T_B}{ln\left(\frac{\Delta T_A}{\Delta T_B}\right)}$$

ΔT_A *and* ΔT_B are the temperature differences between fluids at ends A and B. (Assume counterflow operation for multiple pass and crossflow exchangers.)

HEAT EXCHANGERS

$$q = UA\Delta T_m \quad \text{(single pass)}$$

$$q = F_c UA\Delta T_m \quad \text{(multiple pass)}$$

$$\frac{1}{U} \approx \frac{1}{h_o} + \frac{1}{h_i} + R_{fo} + R_{fi} \quad \text{(with fouling factors)}$$

F_c is 1.0 if either fluid is constant in temperature (e.g., changing phase).

STEFAN-BOLTZMAN CONSTANT

$$\sigma = 0.1713 \times 10^{-8} \frac{\text{BTU}}{\text{hr-ft}^2\text{-}^\circ\text{R}^4}$$

RADIATION FROM A GREY BODY

$$E_g = \epsilon_g \sigma T^4$$

ϵ is the emissivity.
T is in °R.

NET RADIATION TRANSFER BETWEEN TWO GREY BODIES

$$E = \sigma F_e F_a [(T_1)^4 - (T_2)^4]$$
$$= \sigma F_{12} [(T_1)^4 - (T_2)^4]$$

$F_{12} = \epsilon_{inner}$ for completely enclosed bodies and infinite parallel planes.

HEATING, VENTILATING, AND AIR CONDITIONING

(Q is in ft³/hr. ω is in lbm/lbm. q is in BTU/hr.)

HEATING LOAD

HEAT LOSSES THROUGH WALLS

$$q = UA\,(T_i - T_o)$$

INFILTRATION HEAT LOSSES

$$q_a = 0.24\,Q\rho\,(T_i - T_o)$$
$$\approx 0.018\,Q\,(T_i - T_o)$$

HEAT TO WARM MOISTURE

$$q_w = m_w c_p \Delta T = \rho Q \omega c_p\,(T_i - T_o)$$
$$\approx (0.075)(Q)(\omega)(0.45)(T_i - T_o)$$
$$\approx (0.0338)\omega Q(T_i - T_o)$$

TOTAL HEAT TO WARM INCOMING AIR

$$q_t = q_a + q_w$$
$$= m_a\,(h_i - h_o)$$

LATENT HEAT TO ADD MOISTURE

$$q = h_{fg} Q \rho (\omega_i - \omega_o) \approx 79.5 Q(\omega_i - \omega_o)$$

HEAT CONTRIBUTED BY LIGHTS AND MACHINES

$$q_{\text{motors}} = \frac{(2545)(hp)}{\eta}$$

$$q_{\text{lights}} = (3413)\,(kw)$$

DEGREE DAYS

$$DD = N(65 - T_{\text{ave}})$$

N is the number of days in the heating season.

ANNUAL FUEL CONSUMPTION

$$\text{fuel consumption} = \frac{24q(DD)}{(T_i - T_o)\,(HV)\eta_c}$$

HV is the fuel heating value.

A therm is equivalent to 100,000 BTU/hr.

VENTILATING

15 cfm of fresh air per person is commonly used as a design standard. In areas with excessive activity and smoking, this should be increased to 25-40 cfm.

AIR CONDITIONING PROCESSES

SENSIBLE COOLING

$$q_{\text{removed}} = m_a(h_1 - h_2) \approx m_a(0.24 + 0.45\omega)(T_1 - T_2)$$

$$BF = \text{bypass factor} = \frac{T_{2,db} - T_{\text{coil}}}{T_{1,db} - T_{\text{coil}}}$$

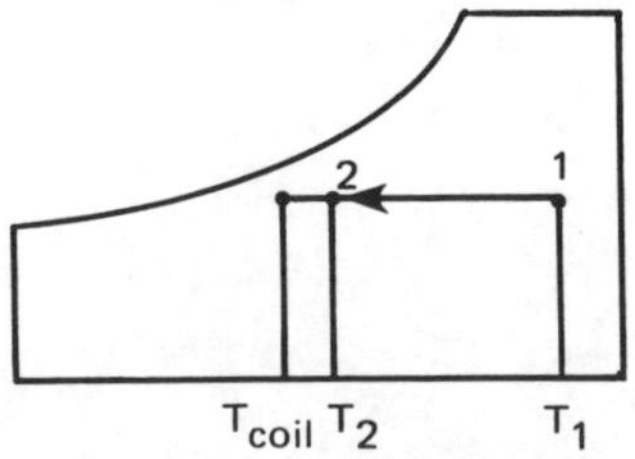

SENSIBLE HEATING

$$q_{\text{added}} = m_a(h_2 - h_1) \approx m_a(0.24 + 0.45\omega)(T_2 - T_1)$$

$$BF = \frac{T_{\text{coil}} - T_2}{T_{\text{coil}} - T_1}$$

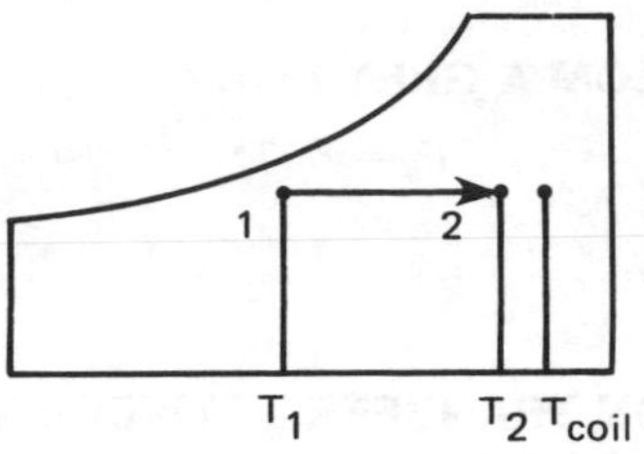

ADIABATIC MIXING OF TWO AIR STREAMS

h and ω are linear scales on the psychrometric chart. Therefore, it is possible to use straight line proportions in the same ratio as m_{a1}/m_{a2}. To do so, draw a straight line between the two points (1 and 2) representing the input streams. The final mixture (3) will be on that line. Use the *lever rule* (on the basis of air masses) to locate the mixture point. Since the water vapor adds little to the mixture volume, the air volumes can be approximated by the total mixture volumes.

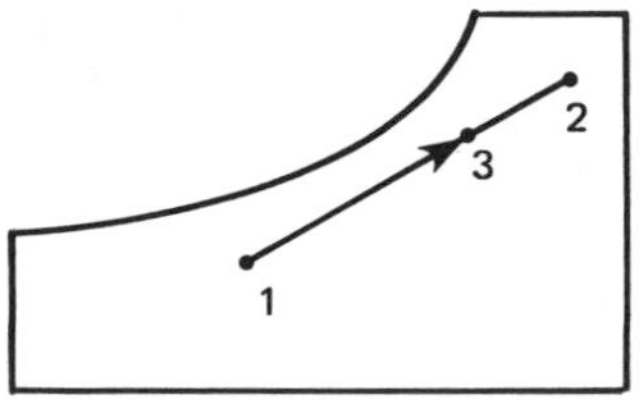

COOLING AND COIL DEHUMIDIFICATION

For convenience, it is assumed that a straight line can be drawn between point 1 and the ADP, and that the final condition of the air will be along that straight line. Point 2 can be located if the coil bypass factor is known.

$$BF = \frac{T_{db,2} - ADP}{T_{db,1} - ADP} = \frac{\text{line segment } 4 - 2}{\text{line segment } 4 - 1}$$

$$q_t = m_a(h_1 - h_2)$$

$$q_l = m_a(\omega_1 - \omega_2)h_{fg}$$

$$q_s = q_t - q_l$$

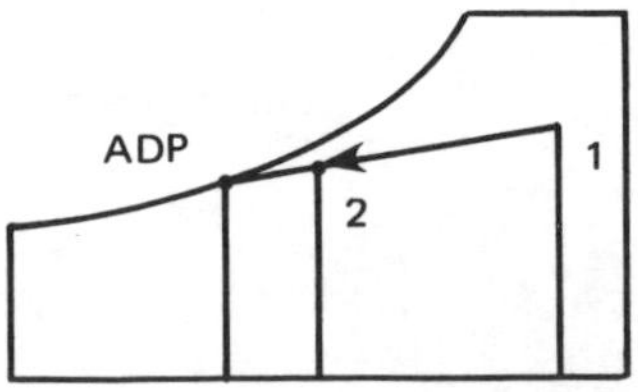

COOLING BY ADIABATIC SATURATION (EVAPORATIVE COOLING)

$$\eta_{sat} = 1 - BF$$
$$= \frac{T_{a,out} - T_{a,in}}{T_{a,in} - T_w}$$

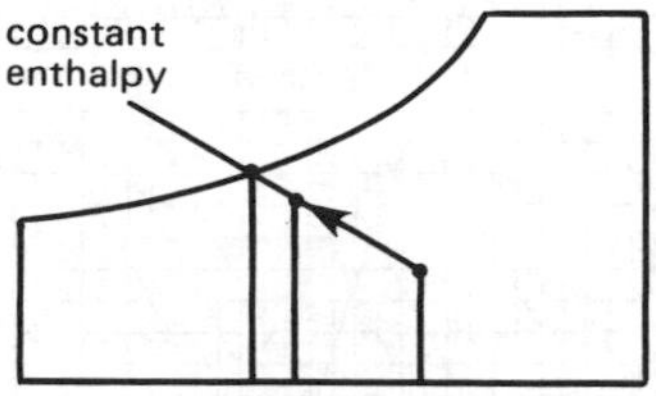

COOLING LOAD

INSTANTANEOUS HEAT GAIN THROUGH WALLS

$$q = UA\Delta T_{te}$$

ΔT_{te} is the total equivalent temperature difference.

HEAT GAIN THROUGH WINDOWS

$$q = A[C_s F_{shg} + U(T_o - T_i)]$$

C_s is the shading coefficient.

F_{shg} is the solar heat gain factor.

HEAT GAIN FROM LIGHTS AND EQUIPMENT

$$q = 3.413\ (P_{watts})$$

$$q = \frac{2545\ (P_{horsepower})}{\eta}$$

ENERGY EFFICIENCY RATIO

$$EER = \frac{\text{cooling (BTU/hr)}}{\text{input power (watts)}}$$

PSYCHROMETRIC CHART

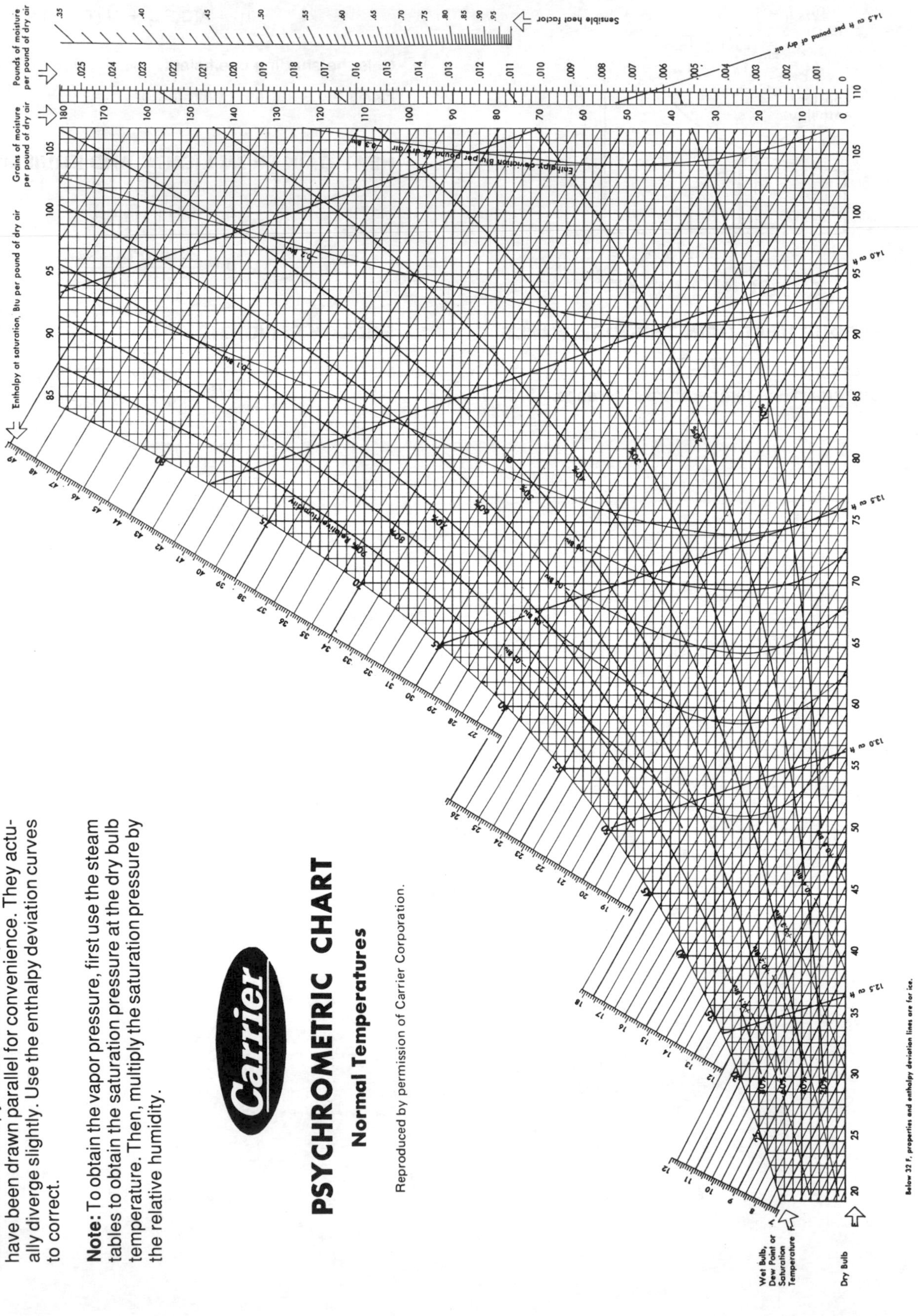

Note: The enthalpy and wet bulb temperature scales have been drawn parallel for convenience. They actually diverge slightly. Use the enthalpy deviation curves to correct.

Note: To obtain the vapor pressure, first use the steam tables to obtain the saturation pressure at the dry bulb temperature. Then, multiply the saturation pressure by the relative humidity.

FORCES

A force is a vector quantity. As such, it is completely defined by giving its magnitude, line of application, sense, and point of application. Units of force are pounds in the English system and newtons in the SI system.

RESULTANTS

The resultant of n forces with components $F_{x,i}$ and $F_{y,i}$ has a magnitude of

$$R = \sqrt{(\Sigma F_{x,i})^2 + (\Sigma F_{y,i})^2}$$

The direction is

$$\phi = \arctan\,[\Sigma F_{y,i}/\Sigma F_{x,i}]$$

COUPLES

A couple is a moment created by two equal forces acting parallel but with opposite directions. The effect of a couple is to cause rotation. If d is the separation distance then

$$M = Fd$$

FORCES IN VECTOR FORM

A force R may be separated into its components by using the direction cosines.

$$F_x = R\cos\theta_x$$

$$F_y = R\cos\theta_y$$

$$F_z = R\cos\theta_z$$

$$\cos\theta_x = x/\sqrt{x^2 + y^2 + z^2}$$

$$\cos\theta_y = y/\sqrt{x^2 + y^2 + z^2}$$

$$\cos\theta_z = z/\sqrt{x^2 + y^2 + z^2}$$

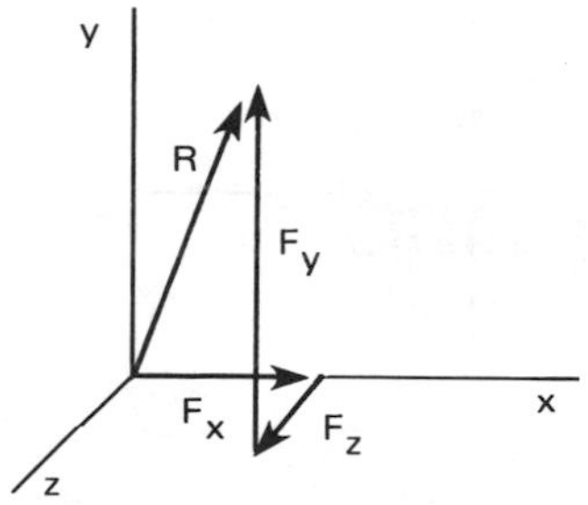

The resultant may be written in vector form in terms of its components and the unit vectors. (The addition is vector addition, not algebraic addition.)

$$\mathbf{R} = \mathbf{i}F_x + \mathbf{j}F_y + \mathbf{k}F_z$$

COMPONENTS OF INCLINED MEMBERS

A non-trigonometric method can be used to resolve forces in inclined members into their components. This resolution can be accomplished by multiplying the force by the ratio of sides, as determined from the geometry of the inclined member.

$$F_y = (y/h)F$$

$$F_x = (x/h)F$$

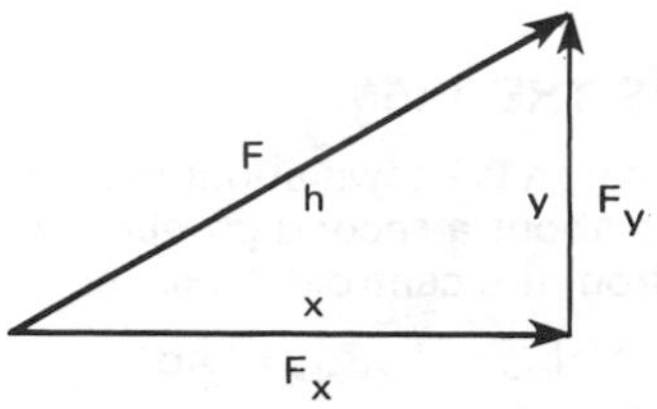

MOMENTS

The moment produced by a force acting with a moment arm of length d is

$$M = Fd$$

CONDITIONS FOR EQUILIBRIUM

In general, the conditions for equilibrium are:

$$\Sigma F_x = 0,\ \Sigma F_y = 0,\ \Sigma F_z = 0,\ \Sigma M_x = 0,\ \Sigma M_y = 0,\ \Sigma M_z = 0$$

For general coplanar systems, the conditions are:

$$\Sigma F_x = 0,\ \Sigma F_y = 0,\ \Sigma M_z = 0$$

For parallel systems, the conditions are:

$$\Sigma F_x = 0,\ \Sigma M_z = 0$$

For concurrent systems, the conditions are:

$$\Sigma F_x = 0,\ \Sigma F_y = 0$$

SIGN CONVENTIONS FOR TRUSS FORCES

For forces in truss members, tension is positive and compression is negative. When freebodies are drawn for truss *joints*, forces leaving the joints place the truss member in tension; forces going into the joints place the member in compression.

DETERMINATE TRUSSES

A truss will be determinate if the number of truss members is equal to

no. truss members = 2(no. of joints) - 3

If the left-hand side is less than the right-hand side, the truss is not rigid. If the left-hand side is greater than the right-hand side, the truss is indeterminate to the degree of the difference.

CATENARY CABLES

In the figure and equations below, c is the distance from the lowest point on the cable to a reference plane below. The distance c is a constant which must be determined. It does not correspond to any physical dimension.

$$y = c\,(\cosh(x/c))$$

$$s = c\,(\sinh(x/c))$$

$$y = \sqrt{s^2 + c^2}$$

$$S = c\,(\cosh\,(a/c) - 1)$$

$$\tan\theta = s/c$$

$H = wc$	(horizontal component of tension)
$F = ws$	(applied vertical load due to cable weight)
$T = wy$	(tangential tension)

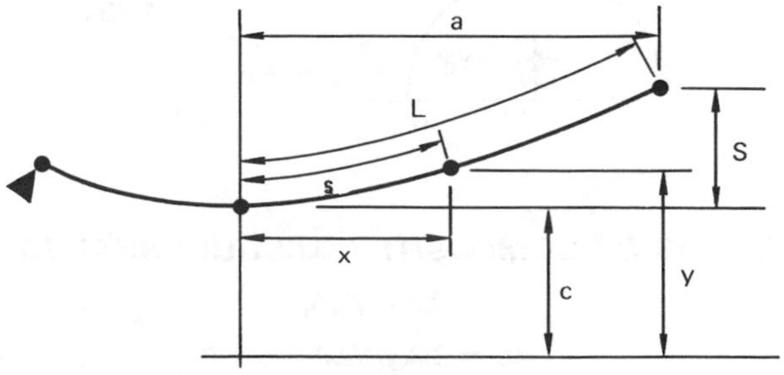

STATICS

FRICTION

The frictional force depends on the coefficient of friction (μ) and the normal force (N).

$$F_f = \mu N \qquad \text{[impending motion]}$$

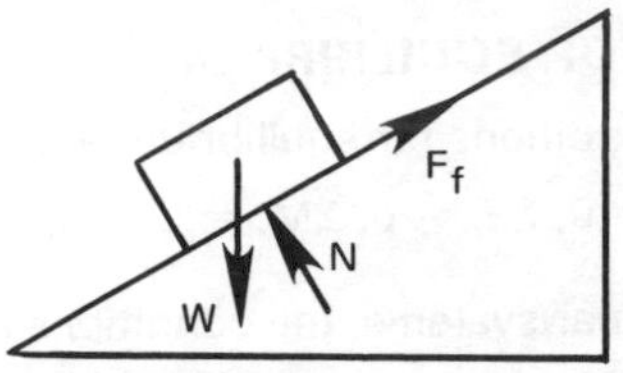

CENTROIDS

The centroid of an object is that point at which the object would balance if suspended. In general, the *x* and *y* coordinates of the centroidal location can be found from the following relationships:

$$x_c = (1/A)\int x\, dA$$

$$y_c = (1/A)\int y\, dA$$

The centroidal locations of common shapes are given below:

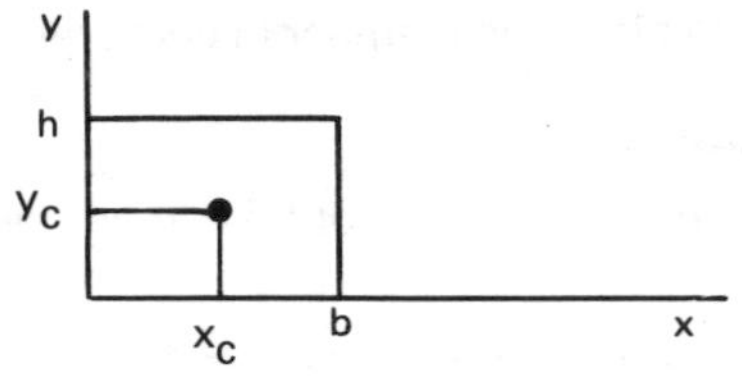

$$x_c = \tfrac{1}{2}b$$

$$y_c = \tfrac{1}{2}h$$

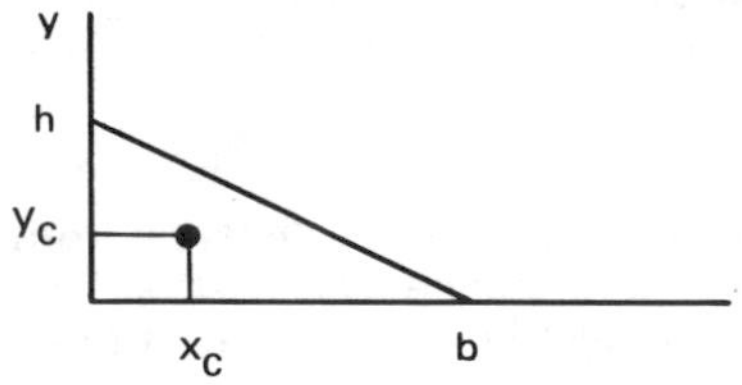

$$x_c = (\tfrac{1}{3})b$$

$$y_c = (\tfrac{1}{3})h$$

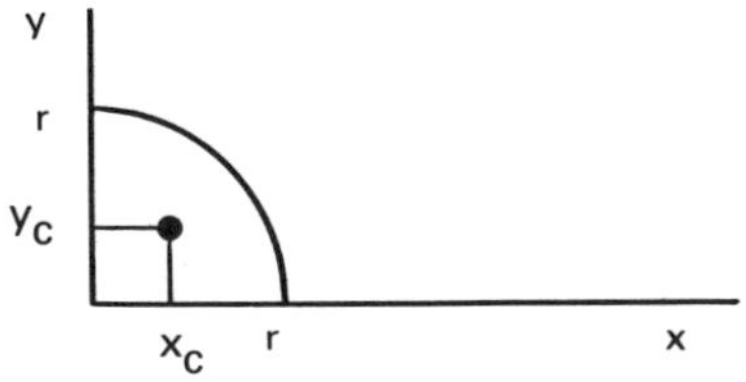

$$x_c = 4r/3\pi$$

$$y_c = 4r/3\pi$$

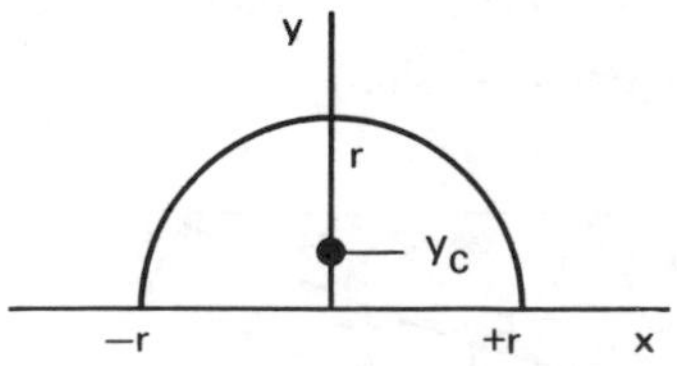

$$x_c = 0$$

$$y_c = 4r/3\pi$$

CENTROIDS OF COMPOSITE (BUILT-UP) SHAPES

$$x_c = \Sigma A_i x_{c,i} / \Sigma A_i$$

$$y_c = \Sigma A_i y_{c,i} / \Sigma A_i$$

MOMENTS OF INERTIA

The moment of inertia (second moment of area) has units of (length)^4 and can be considered a measure of resistance to bending. I_x and I_y represent the resistance to bending about the *x* and *y* axes respectively. I_x and I_y are not components.

$$I_x = \int y^2 dA$$

$$I_y = \int x^2 dA$$

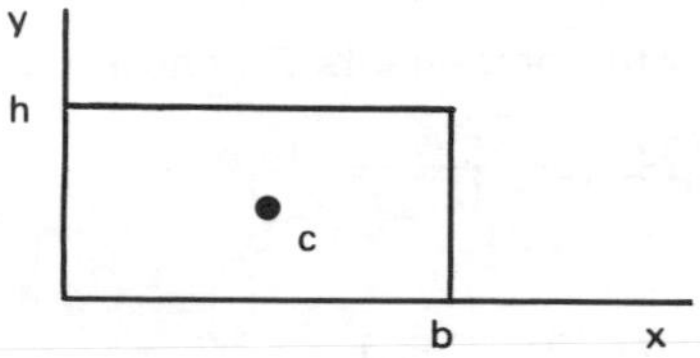

$$I_{x,c} = (1/12)bh^3$$

$$I_x = (1/3)bh^3$$

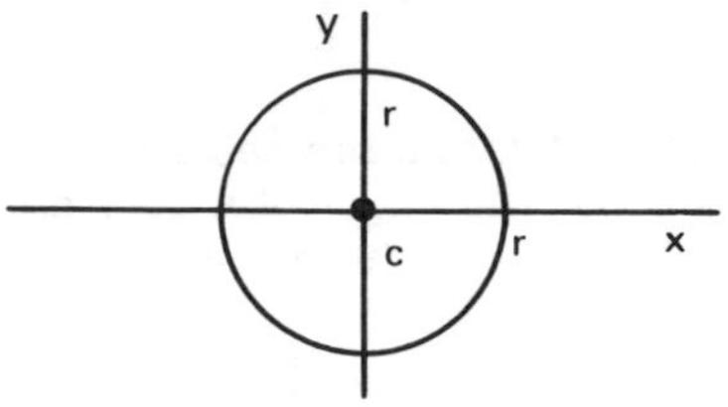

$$I_{x,c} = \tfrac{1}{4}\pi r^4$$

RADIUS OF GYRATION

The radius of gyration is the distance from a reference axis at which all of the area can be considered to be concentrated to produce the actual moment of inertia.

$$k = \sqrt{I/A} \text{ or } k = \sqrt{J/A} \text{ for polar moments of inertia}$$

POLAR MOMENTS OF INERTIA

In general, the polar moment of inertia is

$$J = \int r^2 dA = I_x + I_y$$

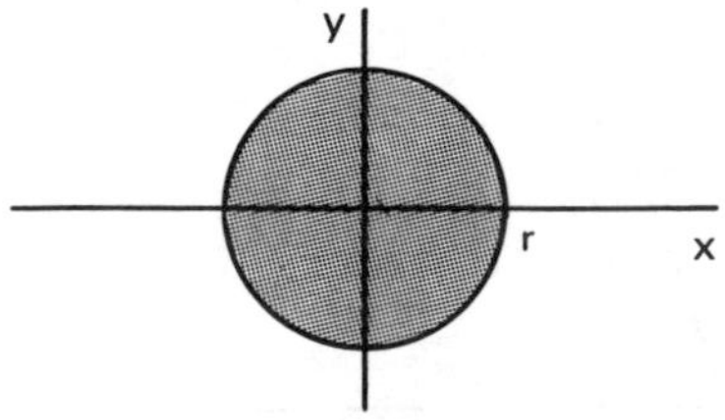

$$J = \tfrac{1}{2}\pi r^4 = (1/32)\pi d^4$$

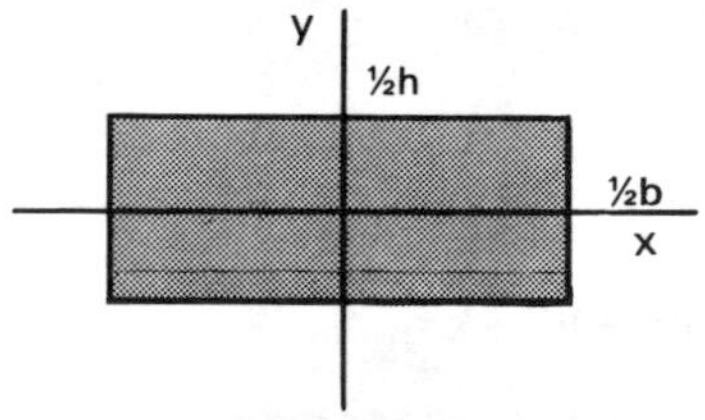

$$J = (1/12)bh(b^2 + h^2)$$

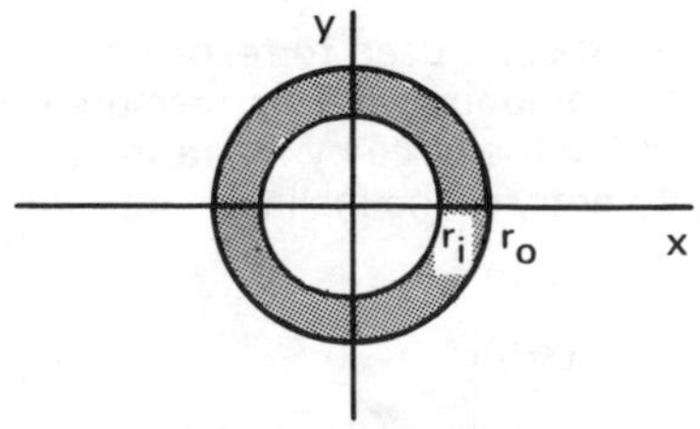

$$J = \tfrac{1}{2}\pi(r_o^4 - r_i^4)$$

PARALLEL AXIS THEOREM

If the moment of inertia is known about the centroidal axis, the moment of inertia about a second parallel axis located a distance of *d* away from the centroidal axis is:

$$I_{new} = I_{centroidal} + Ad^2$$

TENSILE TEST

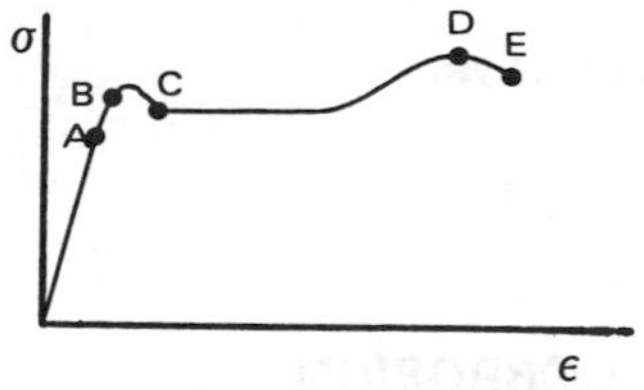

engineering stress: $\sigma = P/A_o$

true stress: $\Sigma = P/A_{instantaneous}$

engineering strain: $\epsilon = \Delta L/L_o$

true strain: $E = \ln(A_o/A_{instantaneous})$

point A: proportionality limit - the highest stress for which Hooke's law ($\sigma = E\epsilon$) is valid.

point B: elastic limit - the highest stress for which no permanent deformation occurs.

point C: yield point - the stress at which a sharp drop in load-carrying ability occurs.

point D: ultimate strength - the highest stress which the material can achieve.

point E: fracture strength - the stress at fracture.

The *modulus of elasticity* (E) is the slope of the line for stresses up to the proportionality limit.

The *shear modulus* (G) can be calculated from the modulus of elasticity and Poisson's ratio.

$$G = E/2(1 + \mu)$$

The *toughness* is the work per unit volume required to cause fracture. It is calculated as the area under the σ-ϵ curve up to the point of fracture. Units of toughness are in-lbf/in^3.

The *ductility* is a measure of the amount of plastic strain at the breaking point. It is calculated as the per cent reduction in area at fracture.

$$\text{reduction in area} = (A_o - A_{frac.})/A_o = (L_{frac.} - L_o)/L_o$$

The *per cent elongation at fracture* is calculated from the original length and the fracture length after the sample has 'snapped back.'

ENDURANCE TEST

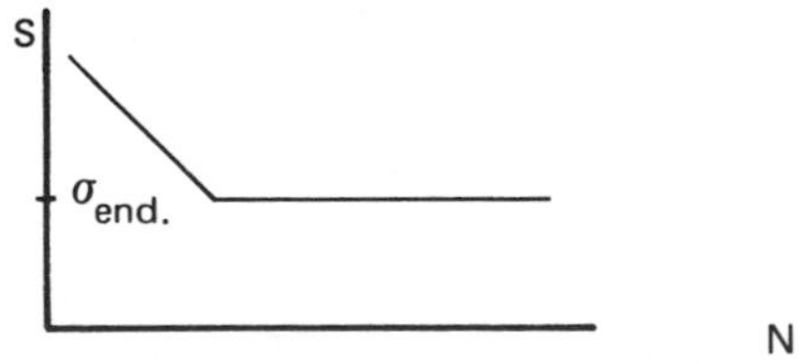

Endurance tests (fatigue tests) apply a cyclical loading of constant maximum amplitude. The plot (usually semi-log or log-log) of the maximum stress and the number of cycles to failure is known as an S-N plot.

The *endurance stress (endurance limit or fatigue limit)* is the maximum stress which can be repeated indefinitely without causing failure.

The *fatigue life* is the number of cycles required to cause failure for a given stress level.

IMPACT TEST

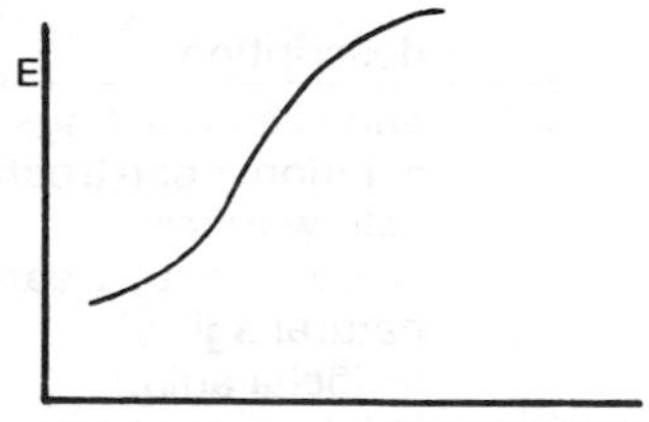

Impact tests determine the amount of energy required to cause failure in standardized test samples. The tests are repeated over a range of temperatures to determine the *transition temperature.* (The transition temperature is approximately 32°F for low-carbon steel.)

CREEP TEST

A constant stress less than the yield strength is applied and the elongation versus time is measured.

The *creep rate* ($d\epsilon/dt$) is very temperature dependent.

The *creep strength* is the stress which results in a given creep rate.

The *rupture strength* is the stress which results in failure after some given amount of time.

HARDNESS VERSUS ULTIMATE STRENGTH

$$S_{ut} \approx (500)\,(BHN)$$

STEEL ALLOYING INGREDIENTS

(XX is the carbon content, 0.XX%)

alloy number	major alloying elements
10XX	plain carbon steel
11XX	resulfurized plain carbon
13XX	manganese
23XX, 25XX	nickel
31XX, 33XX	nickel, chromium
40XX	molybdenum
41XX	chromium, molybdenum
43XX	nickel, chromium, molybdenum
46XX, 48XX	nickel, molybdenum
51XX	chromium
61XX	chromium, vanadium
81XX, 86XX, 87XX	nickel, chromium, molybdenum
92XX	silicon

ALUMINUM ALLOYING INGREDIENTS

alloy number	major alloying ingredient
1XXX	commercially pure aluminum (99+%)
2XXX	copper
3XXX	manganese
4XXX	silicon
5XXX	magnesium
6XXX	magnesium and silicon
7XXX	zinc
8XXX	other

MATERIALS SCIENCE

ALUMINUM TEMPERS

temper	description
T2	annealed (castings only)
T3	solution heat-treated, followed by cold working
T4	solution heat-treated, followed by natural aging
T5	artificial aging
T6	solution heat-treated, followed by artificial aging
T7	solution heat-treated, followed by stabilizing by overaging heat treating
T8	solution heat-treated, followed by cold working and subsequent artificial aging

GALVANIC SERIES

(Anodic to Cathodic)

Magnesium alloys
Alclad 3S
Aluminum alloys
Low-carbon steel
Cast iron
Stainless—No. 410
Stainless—No. 430
Stainless—No. 404
Stainless—No. 316
Hastelloy A
Lead-tin alloys
Brass
Copper
Bronze
90/10 Copper-nickel
70/30 Copper-nickel
Inconel
Silver
Stainless steels (passive)
Monel
Hastelloy C
Titanium

PREVENTING CORROSION

In many designs, corrosion can be reduced or eliminated entirely by avoiding conditions conducive to corrosion. Several design principles for avoiding corrosion are available.

- Metals should be chosen on the basis of their potential for corrosion. This requires attention to chemical properties and environment. When two different metals must be in contact, they should be as close together in the galvanic series as possible.
- Protective coatings (e.g., sodium silicate, sodium benzoate, and various organic amines) can be used.
- Dampness should be eliminated. If there is no electrolyte, there can be no corrosion.
- Since each metal exhibits maximum corrosion at a particular pH, acidity or alkalinity of the environment can be controlled.
- An electrical current can be applied to the corrosion circuit to counter the corrosion reaction. This is known as *cathodic protection.*
- *Sacrificial anodes* made from active metals (e.g., magnesium or calcium) can be placed in areas where corrosion would otherwise occur. These anodes intercept the corrosive current (i.e., they give off electrons) and are sacrificed in the process of saving the structure.

STRESS

Normal stress: $\sigma = \frac{F}{A}$

Shear stress: $\tau = \frac{F}{A}$

STRAIN

$$\epsilon = \frac{\Delta L}{L_o}$$

HOOKE'S LAW

For normal stress: $\sigma = E\epsilon$

For shear stress: $\tau = G\phi$ [ϕ in radians]

ELONGATION UNDER NORMAL STRESS

$$\delta = L_o\epsilon = \frac{FL_o}{AE}$$

VALUES OF E AND G

Steel: $E = 3 \times 10^7$ psi (hard), 2.9×10^7 psi (soft);
$G = 1.15 \times 10^7$ psi

Alumininum: $E = 1 \times 10^7$ psi; $G = 3.85 \times 10^6$ psi

POISSON'S RATIO

μ is the ratio of traverse strain to longitudinal strain.

$$\mu = \frac{\Delta D L_o}{D_o \Delta L}$$

Typical values of μ are .3 for steel and .33 for aluminum.

THERMAL STRESS AND STRAIN

The elongation of an object when heated is

$$\Delta L = \alpha L_o \Delta T$$

Values of α are 6×10^{-6} in/in-°F for steel and 1.3×10^{-5} in/in-°F for aluminum.

The thermal stress and strain are:

$$\epsilon_{th} = \alpha\Delta T$$

$$\sigma_{th} = \epsilon_{th} E$$

NORMAL STRESS IN A BEAM

Normal stress in a beam is also called 'flexure stress' and 'bending stress'.

$$\sigma = \frac{My}{I}$$

$$\sigma_{max} = \frac{Mc}{I} = \frac{M}{Z}$$

$$I = \frac{bh^3}{12} \quad \text{(for a rectangle)}$$

$$Z = \text{section modulus} = \frac{I}{c}$$

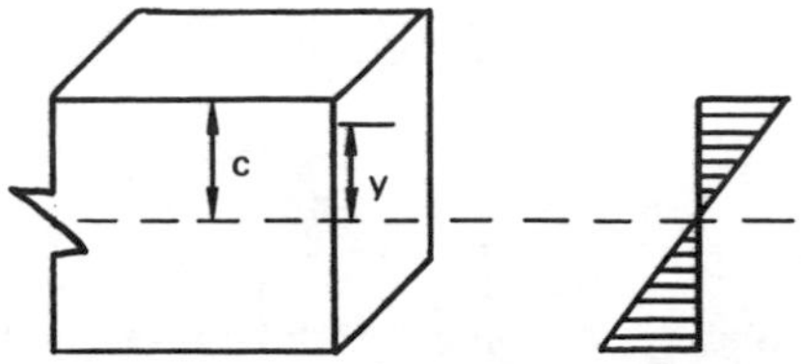

SHEAR STRESS IN A BEAM

$$\tau = \frac{VQ}{Ib}$$

$$\tau_{max} = \frac{3V}{2A} \quad \text{(rectangle)}$$

$$= \frac{4V}{3A} \quad \text{(circular)}$$

V = shear (pounds)

Q = statical moment, also known as 'first moment of the area'

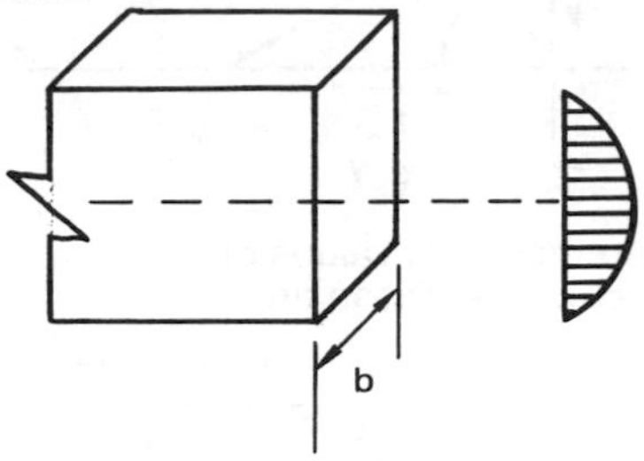

TORSIONAL STRESS AND STRAIN IN A SHAFT

$$\tau = G\phi = \frac{Tr}{J}$$

J = polar (area) moment of inertia

$= \frac{1}{2}\pi r^4$ (round shafts)

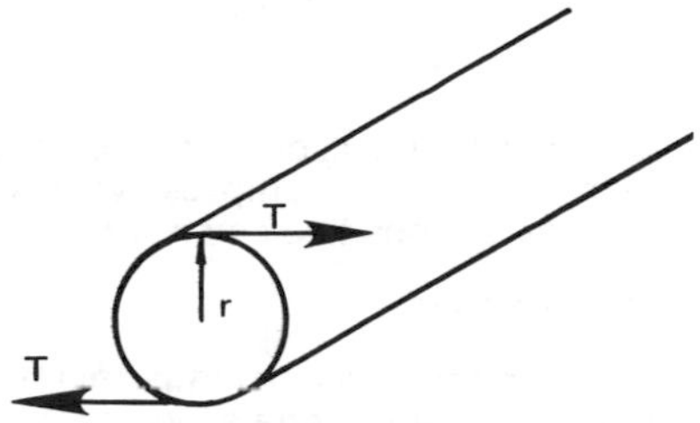

$$\phi = \text{angle of twist} = \frac{\tau}{G} = \frac{TL}{GJ} \quad \text{(in radians)}$$

The maximum torque that the shaft can carry is

$$T_{max} = \frac{\tau_{max} J}{r}$$

The horsepower transmitted by the shaft is

$$P_{hp} = \frac{2\pi T_{ft\text{-}lbf}(rpm)}{33,000}$$

$$P_{hp} = \frac{T_{in\text{-}lbf}(rpm)}{63,025}$$

ECCENTRIC NORMAL STRESS

$$\sigma_{max} = \frac{F}{A} \pm \frac{Mc}{I}$$

$$M = Fe$$

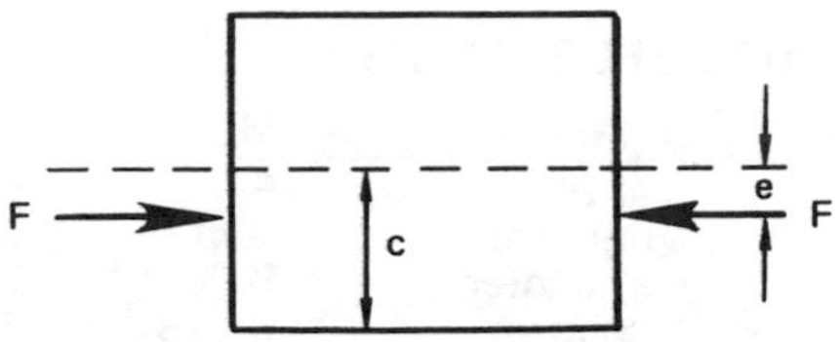

COMBINED STRESS

The normal and shear stresses on a plane whose normal is inclined an angle θ from the horizontal are

$$\sigma_\theta = \tfrac{1}{2}(\sigma_x + \sigma_y) + \tfrac{1}{2}(\sigma_x - \sigma_y)\cos 2\theta + \tau \sin 2\theta$$

$$\tau_\theta = -\tfrac{1}{2}(\sigma_x - \sigma_y)\sin 2\theta + \tau \cos 2\theta$$

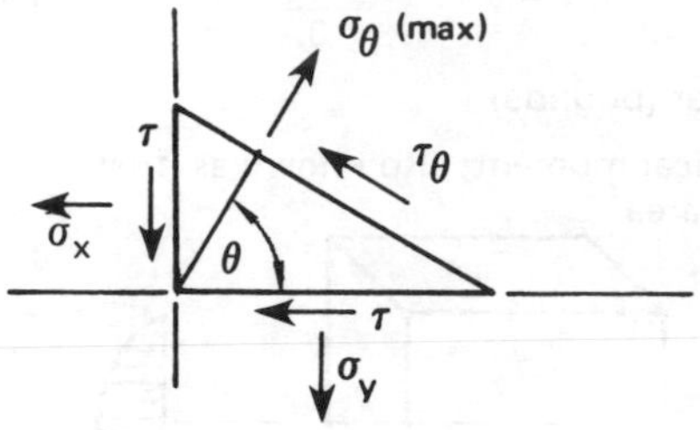

The maximum and minimum values of σ_θ and τ_θ (as θ is varied) are the principal stresses. These are

$$\sigma(\text{max, min}) = \tfrac{1}{2}(\sigma_x + \sigma_y) \pm \tau(\text{max})$$

$$\tau(\text{max, min}) = \pm \tfrac{1}{2}\sqrt{(\sigma_x - \sigma_y)^2 + (2\tau)^2}$$

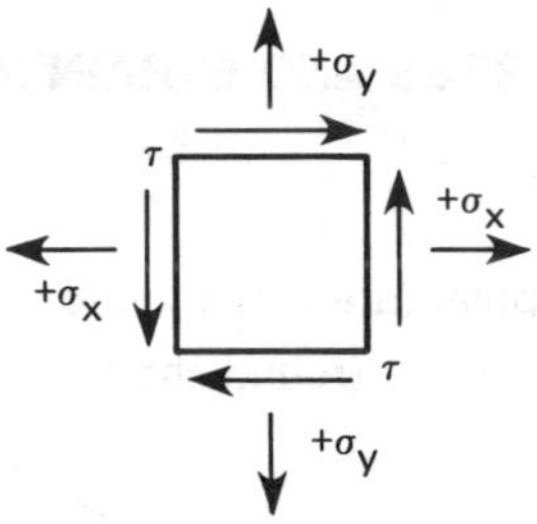

Proper sign convention must be adhered to. Normal tensile stresses are positive; normal compressive stresses are negative. Shear stresses are positive as shown.

ALLOWABLE STRESS

The allowable stress may be calculated from either the yield stress (typical for ductile materials like steel) or the ultimate strength (typical for brittle materials like cast iron).

$$\sigma_{\text{allowable}} = S_{\text{yield}}/(FS) \qquad \text{(ductile)}$$

$$= S_{\text{ultimate}}/(FS) \qquad \text{(brittle)}$$

MOMENT DIAGRAMS

- Clockwise moments are positive. (Use the left-hand rule.)
- Concentrated loads produce straight inclined lines.
- Uniform loads produce parabolic lines.
- Maximum moment occurs where shear (V) is zero.
- Moment is zero at a free end or hinge.
- Moment at any point is the area under the shear diagram up to that point. That is, $M = \int V\,dx$.

SHEAR DIAGRAMS

- Loads and reactions acting up are positive.
- Shear at any point is the sum of forces up to that point.
- Concentrated loads produce horizontal straight lines.
- Uniform loads product straight inclined lines.
- Shear at any point is the slope of the moment diagram at that point. That is, $V = dM/dx$.

SIMPLE BEAM DEFLECTIONS

type of beam	loading	M_{max}	deflection
Cantilever	at tip	FL	$FL^3/3EI$
Cantilever	uniform	$\tfrac{1}{2}wL^2$	$wL^4/8EI$
Simple	at center	$\tfrac{1}{4}FL$	$FL^3/48EI$
Simple	uniform	$wL^2/8$	$5wL^4/384EI$

SLENDER COLUMNS IN COMPRESSION

Euler's formula may be used if the actual stress is kept less than the yield strength. Slender columns fail by buckling in a smooth curve. A slender column has a slenderness ratio of approximately 80 or higher.

$$\text{slenderness ratio} = \frac{L}{k}$$

$$F_{\text{buckling}} = \frac{\pi^2 EI}{(CL)^2(FS)}$$

Illus.	end conditions	ideal C	design C
(a)	both ends pinned	1	1.0*
(b)	both ends built in	0.5	0.65*–0.90
(c)	one end pinned, one end built in	0.707	0.80*–0.90
(d)	one end built in, one end free	2	2.0–2.1*
(e)	one end built in, one end fixed against rotation but free	1	1.2*
(f)	one end pinned, one end fixed against rotation but free	2	2.0*

* AISC Manual of Steel Construction values

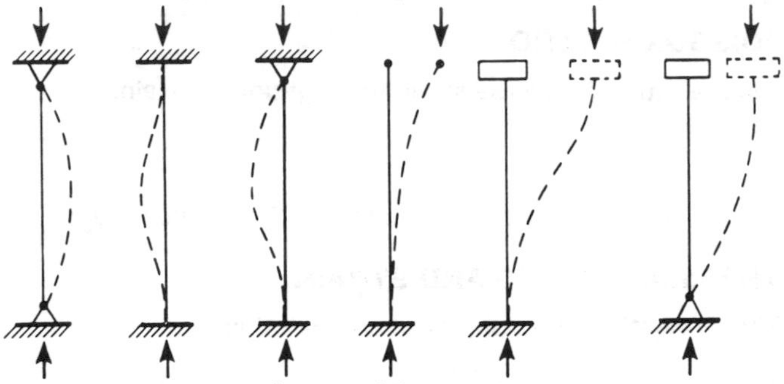

SPRINGS

$$F = kx$$

$$W = \tfrac{1}{2}kx^2$$

$$k = k_1 + k_2 + k_3 + \ldots \qquad \text{(parallel springs)}$$

$$\frac{1}{k} = \frac{1}{k_1} + \frac{1}{k_2} + \frac{1}{k_3} + \cdots \qquad \text{(series springs)}$$

THIN-WALLED CYLINDERS

A tank has thin walls if the wall thickness is less than (1/10) of the tank diameter.

$$\sigma_{\text{hoop}} = \frac{pr}{t}$$

$$\sigma_{\text{long}} = \frac{pr}{2t}$$

The hoop and long stresses are the principal stresses. They do not combine.

For spherical tanks or spherical ends on a cylindrical tank, use the long stress formula.

SHAFTS

$$\tau = \frac{Tr}{J}$$

$$J = \frac{\pi r^4}{2} = \frac{\pi D^4}{32} \qquad \text{(round shafts)}$$

$$J = \frac{\pi}{2}\left[r_o^4 - r_i^4\right] \qquad \text{(hollow shafts)}$$

The angle of twist in radians is

$$\phi = \frac{TL}{GJ}$$

The shear modulus, if unknown, can be calculated from the modulus of elasticity and Poisson's ratio.

$$G = \frac{E}{2(1+\mu)}$$

Horsepower, torque, and rpm are related.

$$T_{\text{in-lbf}} = \frac{(63{,}025)\,(\text{horsepower})}{\text{rpm}}$$

THICK-WALLED CYLINDERS

The circumferential (tangential) and radial stresses are maximum and minimum values at either the inner or outer surfaces.

stress	external pressure, p	internal pressure, p
σ_{co}	$\frac{-(r_o^2+r_i^2)p}{r_o^2-r_i^2}$	$\frac{2r_i^2p}{r_o^2-r_i^2}$
σ_{ro}	$-p$	0
σ_{ci}	$\frac{-2r_o^2p}{r_o^2-r_i^2}$	$\frac{(r_o^2+r_i^2)p}{r_o^2-r_i^2}$
σ_{ri}	0	$-p$
τ_{max}	$(\frac{1}{2})\,\sigma_{ci}$	$(\frac{1}{2})\,(\sigma_{ci}+p)$

The *diametral strain*, $\Delta D/D$, and the *circumferential strain*, $\Delta C/C$, are equal in a circular cylinder under pressure loading.

$$\frac{\Delta D}{D} = \frac{\Delta C}{C} = \frac{\Delta r}{r} = \frac{\sigma_c - \mu(\sigma_r + \sigma_L)}{E}$$

If two cylinders are pressed together with an initial interference, I, the pressure, p, acting between them expands the outer cylinder and compresses the inner one. *Interference* usually means diametral interference.

$$I = |\Delta D_i|_{\text{outer cylinder}} + |\Delta D_o|_{\text{inner cylinder}}$$

When pieces are pressed together, the assembly force can be calculated as a sliding frictional force based on the normal force.

$$F_{\text{max, assembly}} = fN = 2\pi r_{\text{shaft}} Lpf$$

The coefficient of friction for press fits is highly variable, having been reported in the range of .03 to .33.

If the hub is acted upon by a torque or a torque-causing force, the maximum resisting torque is

$$T_{\text{max}} = 2\pi r_{\text{shaft}}^2 Lpf$$

MAXIMUM SHEAR STRESS THEORY (DUCTILE MATERIALS)

Failure occurs when

$$\tau_{max} > \frac{S_{yt}}{2}$$

The factor of safety is

$$FS = \frac{S_{yt}}{2\tau_{max}}$$

The yield point in shear is predicted as 0.5 S_{yt}.

DISTORTION ENERGY AND VON MISES THEORY (DUCTILE MATERIALS)

Failure is predicted as the Von Mises stress.

$$\sigma' > S_{yt}$$

$$\sigma' = \sqrt{\sigma_1^2 + \sigma_2^2 - \sigma_1\sigma_2}$$

σ_1 and σ_2 are the principal stresses

$$FS = \frac{S_{yt}}{\sigma'}$$

The yield point in shear is 0.577 S_{yt}.

ENDURANCE STRENGTH

For steel, the endurance strength is approximately

$$S'_{e,\,steel} = 0.5S_{ut} \qquad [S_{ut} < 200{,}000 \text{ psi}]$$
$$= 100{,}000 \text{ psi} \qquad [S_{ut} > 200{,}000 \text{ psi}]$$

If necessary, the ultimate strength of steel can be calculated from its *Brinell hardness.*

$$S_{ut} \approx (500)(BHN)$$

FLUCTUATING STRESSES

For a part subjected to a fluctuating load, failure cannot be determined solely by yield strength or endurance limit. The combined effects of loading must be considered. The stress is described graphically on a diagram which plots the mean stress versus the alternating stress. Both of these stresses may be either normal stresses or shear stresses. A criterion for failure is established by relating the yield strength, the ultimate strength, and the endurance limit. One method of relating this information is a *Soderberg line.*

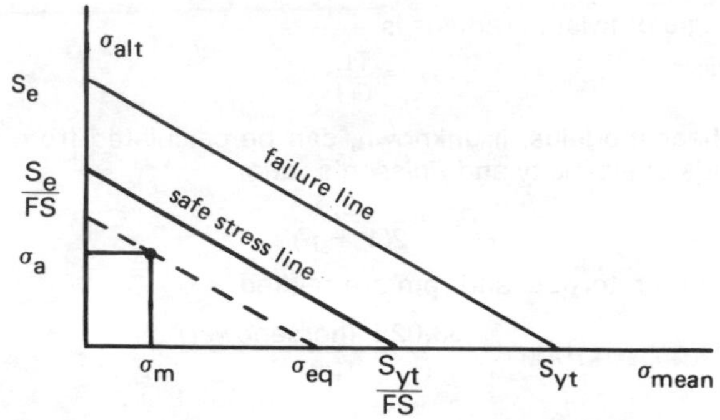

The *mean stress* is

$$\sigma_m = \frac{\sigma_{max} + \sigma_{min}}{2}$$

The *alternating stress* is half the *range stress.*

$$\sigma_a = \frac{\sigma_{max} - \sigma_{min}}{2}$$

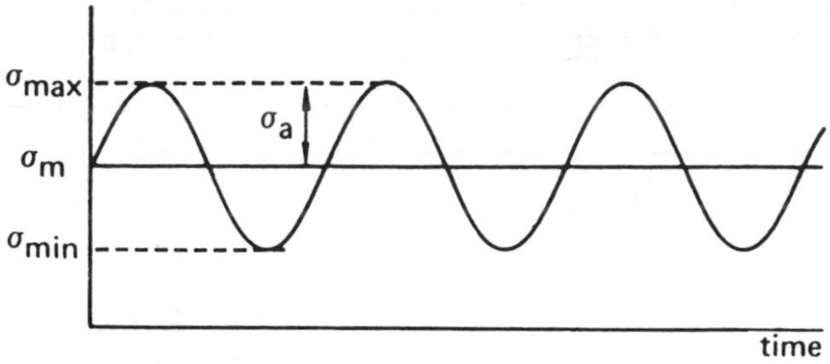

Stress concentration factors (with the notable exception of the Wahl correction factor for springs) are applied to the alternating stress only.

The factor of safety is

$$FS = \frac{S_{yt}}{\sigma_{eq}}$$

CUMULATIVE FATIGUE

The *Palmgren-Miner cycle ratio summation theory* (usually called just *Miner's rule*) often is used to evaluate cumulative damage. If a machine part is subjected to $\sigma_{max,\,1}$ for n_1 cycles, $\sigma_{max,\,2}$ for n_2 cycles, etc., the part should not fail if the equation holds. (N_i are the fatigue lives for the various stress levels.)

$$\sum \frac{n_i}{N_i} < C$$

A value of 1.0 commonly is used for C.

NEWTON'S LAWS

The first law: The velocity (momentum) of an object will not change unless it is acted upon by a force.

The second law: $F = \frac{m_{lbm}a}{g_c}$ or $T = \frac{I_{ft^2\text{-}lbm}\alpha}{g_c}$

The third law: For every action there is an equal but opposite reaction.

Law of Universal Gravitation: The gravitational force between two objects with masses m_1 and m_2 is

$$F = \frac{Gm_1m_2}{d^2}$$

$$G = 3.44 \times 10^{-8} \frac{ft^4}{lbf\text{-}sec^4}$$

DISTANCE, VELOCITY, AND ACCELERATION

$$a = \frac{dv}{dt} = \frac{d^2s}{dt^2}$$

$$v = \frac{ds}{dt} = \int a\, dt$$

$$s = \int v\, dt = \iint a\, dt$$

UNIFORM ACCELERATION FORMULAS

to find	*given these*	*use this formula*
t	a, v_o, v	$t = (v - v_o)/a$
t	a, v_o, s	$t = \frac{\sqrt{2as + v_o^2} - v_o}{a}$
t	v_o, v, s	$t = 2s/(v_o + v)$
a	t, v_o, v	$a = (v - v_o)/t$
a	t, v_o, s	$a = (2s - 2v_ot)/t^2$
a	v_o, v, s	$a = (v^2 - v_o^2)/2s$
v_o	t, a, v	$v_o = v - at$
v_o	t, a, s	$v_o = (s/t) - \frac{1}{2}at$
v_o	a, v, s	$v_o = \sqrt{v^2 - 2as}$
v	t, a, v_o	$v = v_o + at$
v	a, v_o, s	$v = \sqrt{v_o^2 + 2as}$
s	t, a, v_o	$s = v_ot + \frac{1}{2}at^2$
s	a, v_o, v	$s = (v^2 - v_o^2)/2a$
s	t, v_o, v	$s = \frac{1}{2}t(v_o + v)$

ACCELERATION DUE TO GRAVITY

Falling body problems may be solved with the uniform acceleration formulas by substituting $a = g = 32.2\ ft/sec^2 (9.807\ m/s^2)$.

PROJECTILE MOTION

The formulas neglect air drag. Range (r) is maximum when $\phi = 45°$.

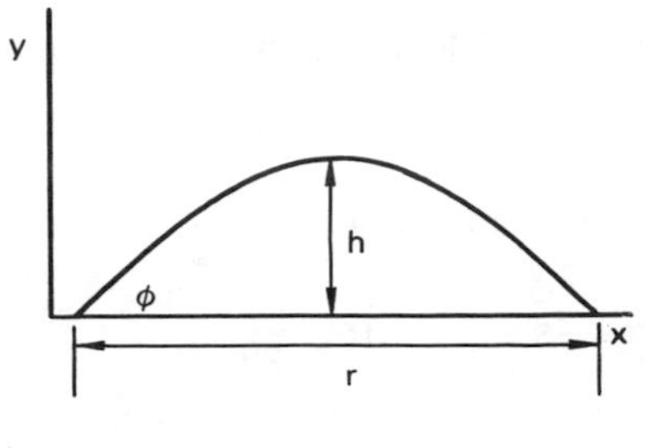

$$x = (v_o\cos\phi)t$$

$$y = (v_o\sin\phi)t - \frac{1}{2}gt^2$$

$$v = \sqrt{v_o^2 - 2gy}$$

$$v_x = v_o\cos\phi$$

$$v_y = v_o\sin\phi - gt$$

$$h = \frac{v_o^2\sin^2\phi}{2g}$$

$$r = \frac{v_o^2\sin 2\phi}{g}$$

$$T = \frac{2v_o\sin\phi}{g}$$

WORK AND ENERGY

Work is an energy transfer that occurs when a force is moved through a distance:

$$W = \int F \cdot ds$$

Potential energy is the energy that an object possesses by virtue of its position in a gravitational field:

$$E_p = \frac{1}{2}kx^2 \quad \text{(in ft-lbf)}$$

For a spring (with rate constant k) compressed an amount x:

$$E_p = \frac{1}{2}kx^2$$

Kinetic energy is the energy that an object possesses by virtue of its velocity:

$$E_k = \frac{mv^2}{2g_c} \quad \text{(translation, in ft-lbf)}$$

$$E_k = \frac{I\omega^2}{2g_c} \quad \text{(rotation, in ft-lbf)}$$

THE WORK-ENERGY PRINCIPLE

In the absence of thermal changes, the work done on (or by) a system is equal to its change in energy.

$$W = \Delta E_p + \Delta E_k$$

Power is the amount of work done per unit time.

$$P = \frac{W}{\Delta t}$$

Power can be calculated from force and velocity.

$$P = Fv \quad \text{(translation)}$$

$$P = T\omega \quad \text{(rotation)}$$

HORSEPOWER REQUIRED TO MAINTAIN VELOCITY

For translation: $hp = Fv/550$ (v in fps)

For rotation: $hp = 2\pi Tn/33{,}000$ (T in ft-lbf, n in rpm)

POWER CONVERSIONS

$$1\ hp = 550 \frac{ft\text{-}lbf}{sec} = 33,000 \frac{ft\text{-}lbf}{min}$$

$$= 0.7457\ kW = 0.7068 \frac{BTU}{sec}$$

$$1\ kW = 737.6 \frac{ft\text{-}lbf}{sec} = 44,250 \frac{ft\text{-}lbf}{min}$$

$$= 1.341\ hp = 0.9483 \frac{BTU}{sec}$$

$$1 \frac{BTU}{sec} = 778.17 \frac{ft\text{-}lbf}{sec} = 46,680 \frac{ft\text{-}lbf}{min}$$

$$= 1.415\ hp$$

DYNAMICS

ROTATIONAL MOTION

Analogous variables for rotational motion:

α: rotational acceleration
ω: rotational speed
θ: rotational position

$$\alpha = \frac{d\omega}{dt} = \frac{d^2\theta}{dt^2}$$

$$\omega = \frac{d\theta}{dt} = \int \alpha \, dt$$

$$\theta = \int \omega \, dt = \int\int \alpha \, dt$$

ROTATIONAL MOMENT OF INERTIA

$$I = \frac{mr^2}{2} \quad \text{(solid cylinder)}$$

$$I = \frac{m(r_i^2 + r_o^2)}{2} \quad \text{(hollow cylinder)}$$

RELATIONSHIP BETWEEN ROTATIONAL AND LINEAR MOTION

$$s = r\theta$$

$$v = r\omega$$

$$a = r\alpha$$

CENTRIFUGAL FORCE

$$F_c = \frac{mv^2}{g_c r}$$

HIGHWAY BANKING

The banking angle (superelevation) needed on a circular curve of radius r is

$$\phi = \arctan\left(\frac{v^2}{gr}\right)$$

PARALLEL AXIS THEOREM

$$I_{new} = I_{centroidal} + md^2$$

RADIUS OF GYRATION

$$k = \sqrt{I/m}$$

IMPULSE-MOMENTUM PRINCIPLE

$$F\Delta t = m\Delta v/g_c \quad \text{(translation)}$$

$$T\Delta t = I\Delta\omega v/g_c \quad \text{(rotation)}$$

IMPACTS

The coefficient of restitution (e) is 1 if the collision is perfectly elastic. e is 0 if the collision is completely inelastic (the bodies stick together).

$$e = \frac{\text{relative separation velocity}}{\text{relative approach velocity}} = \frac{-(v_2' - v_1')}{v_2 - v_1}$$

CONSERVATION OF MOMENTUM IN A COLLISION

$$m_1v_1 + m_2v_2 = m_1v_1' + m_2v_2'$$

If $e = 1$, kinetic energy is also conserved.

UNDAMPED, FREE VIBRATIONS (SIMPLE HARMONIC MOTION)

$$f = \frac{\omega}{2\pi} = \frac{1}{T}$$

$$T = \frac{1}{f} = \frac{2\pi}{\omega}$$

$$\omega = 2\pi f = \frac{2\pi}{T}$$

For a simple spring-mass system,

$$\omega = \sqrt{\frac{k}{m} g_c} = \sqrt{\frac{g}{\delta_{static}}}$$

For a simple pendulum consisting of a mass (m) at the end of a massless cord of length L,

$$\omega = \sqrt{\frac{g}{L}}$$

SHAFT VIBRATIONS (CRITICAL SPEEDS)

The critical speed of a shaft can be found from its free lateral vibration frequency. The general equation for the fundamental frequency (critical speed in revolutions per second) is

$$f = \frac{1}{2\pi}\sqrt{\frac{g\Sigma(m_i\delta_i)}{\Sigma(m_i\delta_i^2)}}$$

m_i is the mass of the ith rotating body, and δ_i is the static deflection of the shaft at the ith body.

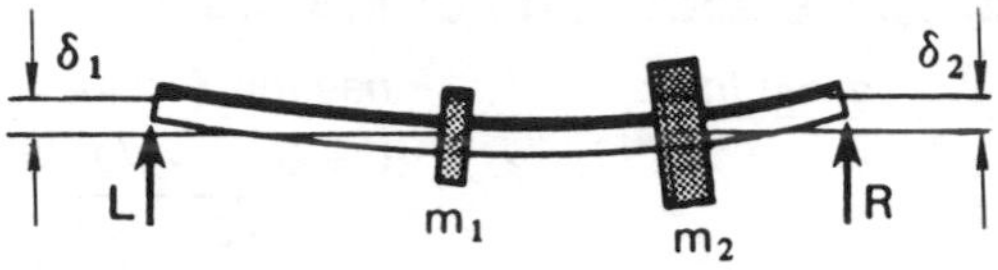

Since this equation is difficult to use, it is common to substitute the *Dunkerly approximation*.

$$\left(\frac{1}{f}\right)^2_{composite} = \left(\frac{1}{f_1}\right)^2 + \left(\frac{1}{f_2}\right)^2 + \left(\frac{1}{f_3}\right)^2 + \cdots$$

The classical solution to the problem of a shaft carrying a single object of weight W assumes a weightless shaft. The deflection at the load is

$$\delta = \frac{Wa^2b^2}{3EIL} \quad \text{(simple supports)}$$

$$\delta = \frac{Wa^3b^3}{3EIL^3} \quad \text{(fixed supports)}$$

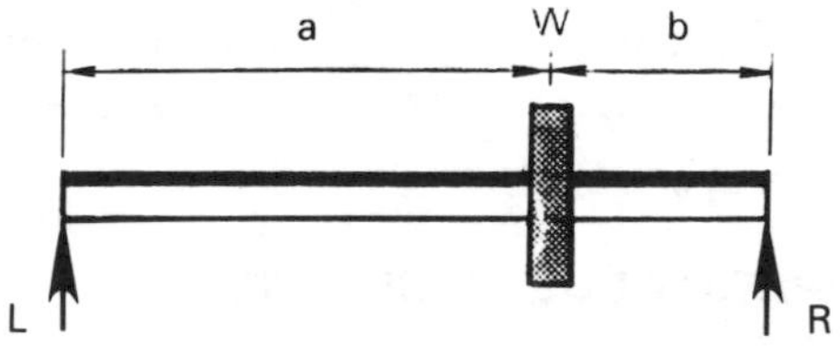

$$I = \frac{\pi r^4}{4} \quad \text{(solid round shafts)}$$

$$f = \frac{1}{2\pi}\sqrt{\frac{g}{\delta}} \quad \text{(consistent units)}$$

DAMPED, FREE VIBRATIONS

C is the damping coefficient, ω is the natural frequency.

$$C_{critical} = \frac{2m\omega}{g_c}$$

$$\text{damping ratio} = \frac{C}{C_{critical}} = \frac{n}{\omega}$$

$$n = \frac{Cg_c}{2m}$$

$n < \omega$ is underdamping

$n = \omega$ is critical damping (fastest return)

$n > \omega$ is overdamping

$$\omega_{damped} = \sqrt{\omega^2 - n^2}$$
$$= \omega\sqrt{1 - \left(\frac{C}{C_{critical}}\right)^2}$$

$$\text{logarithmic decrement} = \frac{2\pi n}{\omega_{damped}}$$

DAMPED, FORCED VIBRATIONS

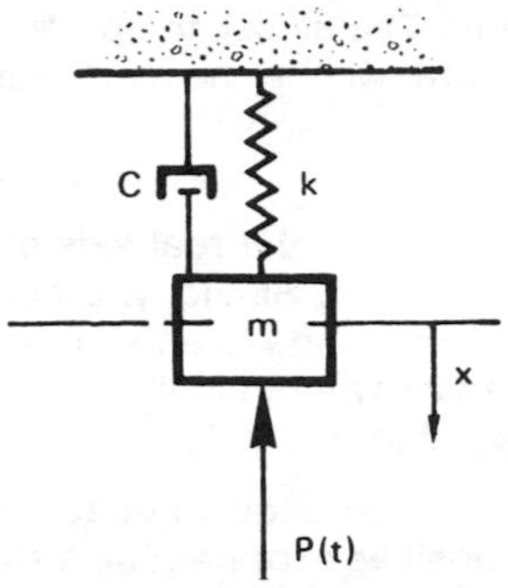

The amplitude of forced vibrations can be found from the pseudo-static deflection and the *magnification factor*, β.

$$A = \beta\left(\frac{P}{k}\right) = \beta\delta_{static}$$

$$\beta = \left|\frac{1}{\sqrt{\left[1 - \left(\frac{\omega_f}{\omega}\right)^2\right]^2 + \left[\frac{g_c C\omega_f}{m(\omega)^2}\right]^2}}\right|$$
$$= \left|\frac{1}{\sqrt{\left[1 - \left(\frac{\omega_f}{\omega}\right)^2\right]^2 + \left[\frac{2C\omega_f}{C_{crit}\omega}\right]^2}}\right|$$

If the forcing frequency, ω_f, is equal to the natural frequency, ω, the magnification factor will be very large. This condition is known as *resonance*.

VIBRATION ISOLATION AND CONTROL

$$f = \frac{1}{2\pi}\sqrt{\frac{g}{\delta_{st}}}$$

$$\sigma_{st} = \frac{\text{equipment weight}}{k}$$

The amount of isolation is known as the *isolation efficiency*.

$$\eta = 1 - \frac{1}{\left(\frac{f_f}{f}\right)^2 - 1}$$

Damping is beneficial only in the case of $(f_f/f) > \sqrt{2}$. Otherwise, damping is detrimental because the transmissibility actually increases above 1.0. The *transmissibility* is the ratio of transmitted force to unbalanced force. For undamped systems ($C = 0$), the transmissibility is equal to the magnification factor, β. For damped systems, the transmissibility is

$$TR = \frac{|P_{transmitted}|}{P_{applied}}$$
$$= \beta\sqrt{1 + \left[\frac{2C\omega_f}{C_{crit}\,\omega}\right]^2}$$

RESPONSE VARIABLES

A dependent variable which predicts the performance of a system is known as a *response variable.* Position, velocity, and acceleration vary with time and are the dependent response variables. Time is usually the independent variable.

NATURAL, FORCED, AND TOTAL RESPONSE

Natural response is induced when energy is applied to an engineering system and subsequently is removed. The system is left alone and is allowed to do what it would do naturally, without the application of further disturbing forces. If a system is acted upon by a force which repeats at regular intervals, the system will move in accordance with that force. This is known as *forced response.* The natural and forced responses are present simultaneously in forced systems. The sum of the two responses is known as the *total response.*

FORCING FUNCTIONS

An equation which describes the introduction of energy into the system as a function of time is known as a *forcing function.* The most common forcing functions used in the analysis of engineering systems are the sine and cosine functions.

$$F(t) = \sin\omega t.$$

TRANSFER FUNCTIONS

The ratio of the system response (output) to the forcing function (input) is known as the *transfer function, T(t).*

$$T(t) = \frac{R(t)}{F(t)}$$

Transfer functions generally are written in terms of the *s* variable. This is accomplished, if *T(t)* is known, by taking the Laplace transform of the transfer function. The result is the *transform of the transfer function,* typically just called the *transfer function.*

$$T(s) = \pounds[T(t)]$$

PREDICTING SYSTEM RESPONSE

Assuming that *T(s)* and $\pounds[F(t)]$ are known, the response function can be found by performing an inverse transformation.

$$R(t) = \pounds^{-1}[R(s)] = \pounds^{-1}[\pounds[F(t)]T(s)]$$

POLES AND ZEROS

A *pole* of the transfer function is a value of *s* which makes *T(s)* infinite. Specifically, a pole is a value of *s* which makes the denominator of *T(s)* zero. A *zero* of the transfer function makes the numerator of *T(s)* zero. Poles and zeros can be real or complex quantities. Poles and zeros can be repeated within a given transfer function—they need not be unique.

A rectangular coordinate system based on the real-imaginary axes is known as an *s-plane.* If poles and zeros are plotted on the s-plane, the result is a *pole-zero diagram.* Poles are represented on the pole-zero diagram as ×'s. Zeros are represented as ○'s.

PREDICTING SYSTEM RESPONSE FROM POLE-ZERO DIAGRAMS

Poles on the pole-zero diagram can be used to predict the usual response of engineering systems. Zeros are not used.

- *pure oscillation:* Sinusoidal oscillation will occur if a pole-pair falls on the imaginary axis. A pole with a value of $\pm ja$ will produce oscillation with a natural frequency of $\omega = a$ radians/sec.
- *exponential decay:* Pure exponential decay is indicated when a pole falls on the real axis. A pole with a value of $-r$ will produce an exponential with time constant $(1/r)$.
- *damped oscillation:* Decaying sinusoids result from pole-pairs in the second and third quadrants of the *s*-plane. A pole-pair having the value $r \pm ja$ will produce oscillation with natural frequency of

  $$\omega_n = \sqrt{r^2 + a^2}$$

  The closer the poles are to the real axis, the greater will be the damping effect. The closer the poles are to the imaginary axis, the greater will be the oscillatory effect.

STABILITY

A pole with a value of $-r$ on the real axis corresponds to an exponential response of e^{-rt}. Similarly, a pole with a value of $+r$ on the real axis corresponds to an exponential response of e^{rt}. However, e^{rt} increases without limit. For that reason, such a pole is said to be unstable.

Since any pole in the first and fourth quadrants of the *s*-plane will correspond to a positive exponential, a stable system must have poles limited to the left half of the *s*-plane (i.e., quadrants two and three).

FEEDBACK THEORY

The basic feedback system consists of a *dynamic unit,* a *feedback element,* a *pick-off point,* and a *summing point.* The summing point is assumed to perform positive addition unless a minus sign is present.

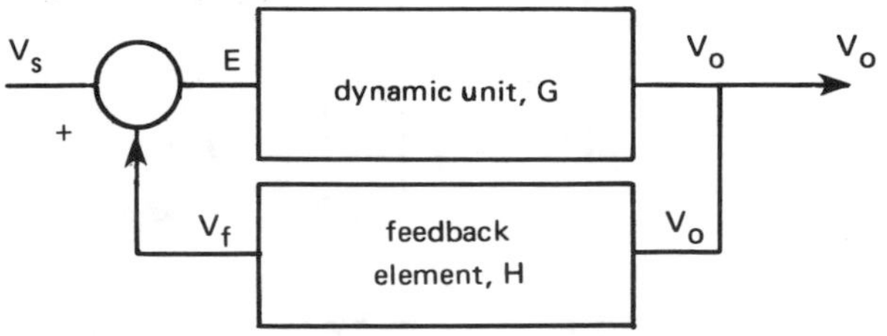

The dynamic unit transforms *E* into V_o according to the *forward transfer function, G.*

$$V_o = GE$$

For amplifiers, the forward transfer function is known as the *direct* or *forward gain.* The difference between the signal and the feedback is known as the *error.*

$$E = V_s + V_f$$

E/V_s is known as the *error ratio.* V_f/V_s is known as the *primary feedback ratio.*

The pick-off point transmits V_o back to the feedback element. The output of the dynamic unit is not reduced by the pick-off point. As the picked-off signal travels through the feedback loop, it is acted upon by the *feedback* or *reverse transfer function, H.*

The output of a feedback system is

$$V_o = GV_s + GHV_o$$

The *closed-loop transfer function* (also known as the *control ratio* or the *system function*) is the ratio of the output to the signal.

$$G_{loop} = \frac{V_o}{V_s} = \frac{G}{1 - GH}$$

STRAIN GAGES

A strain gage is a folded wire which exhibits a resistance change as the wire length changes. The resistance change will be small, and temperature effects should be compensated by using a second unstrained gage as part of the bridge measurement system. Nichrome wire with a total resistance under 1000 ohms commonly is used.

The *strain sensitivity (gage factor)* is defined as

$$K = \left(\frac{\Delta R}{R_o}\right)\left(\frac{L_o}{\Delta L}\right)$$

The strain (generally in microinches per inch) is related to the resistance change.

$$\epsilon = \frac{\Delta R}{K R_o}$$

TEMPERATURE-SENSITIVE RESISTORS

Resistance in conductors will vary with temperature. The change can be positive or negative, and the variation is nonlinear. The variation can be calculated if the coefficients of thermal resistance are known. (β generally is small and is not considered in most analyses).

$$R_T \; + R_o(1 + \alpha\Delta T + \beta\Delta T^2)$$
$$\Delta T = T - T_o$$

Coefficients of Thermal Resistance, α (1/°C)

conductors			
aluminum	0.0043	manganin	0.00002
brass	0.0020	nichrome	0.0004
constantan	0.000002	nickel	0.0068
copper	0.0067	platinum	0.0039
gold	0.0040	silver	0.0041
iron	0.0061	tin	0.0046
lead	0.0039		

ITEM RELIABILITY

Reliability as a function of time, *R(t)*, is the probability that an item will continue to operate satisfactorily up to time *t*. Reliability often is described by the *negative exponential distribution*. Specifically, it is assumed that an item's reliability is

$$R(t) = 1 - F(t) = e^{-\lambda t} = e^{-t/MTBF}$$

This implies that the probability of *x* failures in a period of time is given by Poisson distribution.

$$p\{x\} = \frac{e^{-\lambda}\lambda^x}{x!}$$

The negative exponential distribution is appropriate whenever an item fails only by random causes but never experiences deterioration during its life. This implies that the *expected future life* of an item is independent of the previous duration of operation.

The *hazard function* is defined as the conditional probability of failure in the next time interval given that no failure has occurred thus far. For the exponential distribution, the hazard function is

$$z(t) = \lambda$$

Since this is not a function of *t*, exponential failure rates are not dependent on the length of time previously in operation.

SERIAL SYSTEM RELIABILITY

A serial system will fail if any one of the components fails. The system reliability is

$$R^* = R_1 R_2 R_3 \ldots R_n$$

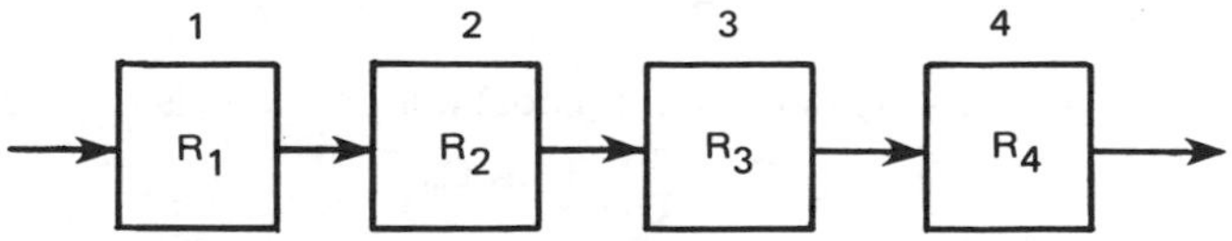

PARALLEL SYSTEM RELIABILITY

A parallel system with *n* items will fail only if all *n* items fail. This property is called *redundancy*, and such a system is said to be redundant.

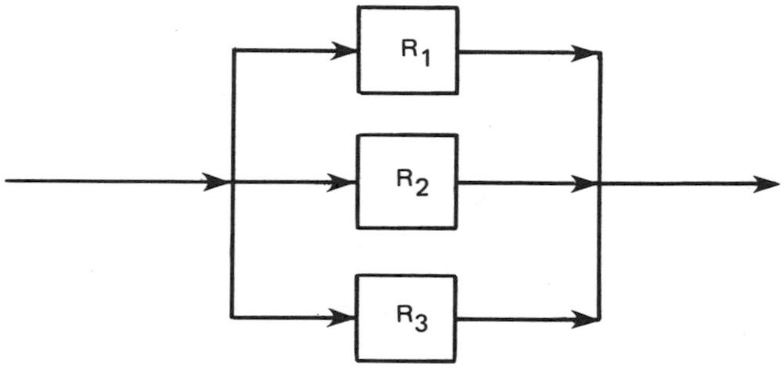

LEARNING CURVES

- T_1 time or cost for the first item
- T_n time or cost for the *n*th item
- n total number of items produced
- b learning curve constant

Learning Curve Constants

learning curve	b
75%	0.415
80%	0.322
85%	0.234
90%	0.152
95%	0.074

The time to produce the *n*th item is given by

$$T_n = T_1(n)^{-b}$$

The total time to produce units from quantity n_1 to n_2 inclusive is

$$\frac{T_1}{1-b}\left[\left(n_2 + \tfrac{1}{2}\right)^{1-b} - \left(n_1 - \tfrac{1}{2}\right)^{1-b}\right]$$

ECONOMIC ORDER QUANTITY

The *economic order quantity* (EOQ) is the order quantity which minimizes the inventory costs per unit time.

- a the constant depletion rate $\left(\frac{\text{items}}{\text{unit time}}\right)$
- h the inventory storage cost $\left(\frac{\$}{\text{item-unit time}}\right)$
- H the total inventory storage cost between orders ($)
- K the fixed cost of placing an order ($)
- Q_0 the order quantity

If the original quantity on hand is Q_0, the stock will be depleted at

$$t^* = \frac{Q_0}{a}$$

The total inventory storage cost between t_o and t^* is

$$H = \tfrac{1}{2}h\frac{Q_0^2}{a}$$

The total inventory and ordering cost per unit time is

$$C_t = \frac{aK}{Q_0} + \tfrac{1}{2}hQ_0$$

C_t can be minimized with respect to Q_o. The EOQ and time between orders are:

$$Q_o^* = \sqrt{2\frac{aK}{h}}$$

$$t^* = \frac{Q_0^*}{a}$$

TABLE OF RELATIVE ATOMIC WEIGHTS

(Based on the atomic mass of $^{12}C = 12$)

name	symbol	atomic number	atomic weight	name	symbol	atomic number	atomic weight
actinium	Ac	89	–	mercury	Hg	80	200.59
aluminum	Al	13	26.9815	molybdenum	Mo	42	95.94
americium	Am	95	–	neodymium	Nd	60	144.24
antimony	Sb	51	121.75	neon	Ne	10	20.183
argon	Ar	18	39.948	neptunium	Np	93	–
arsenic	As	33	74.9216	nickel	Ni	28	58.71
astatine	At	85	–	niobium	Nb	41	92.906
barium	Ba	56	137.34	nitrogen	N	7	14.0067
berkelium	Bk	97	–	nobelium	No	102	–
beryllium	Be	4	9.0122	osmium	Os	76	190.2
bismuth	Bi	83	208.980	oxygen	O	8	15.9994
boron	B	5	10.811	palladium	Pd	46	106.4
bromine	Br	35	79.904	phosphorus	P	15	30.9738
cadmium	Cd	48	112.40	platinum	Pt	78	195.09
calcium	Ca	20	40.08	plutonium	Pu	94	–
californium	Cf	98	–	polonium	Po	84	–
carbon	C	6	12.01115	potassium	K	19	39.102
cerium	Ce	58	140.12	praseodymium	Pr	59	140.907
cesium	Cs	55	132.905	promethium	Pm	61	–
chlorine	Cl	17	35.453	protactinium	Pa	91	–
chromium	Cr	24	51.996	radium	Ra	88	–
cobalt	Co	27	58.9332	radon	Rn	86	–
copper	Cu	29	63.546	rhenium	Re	75	186.2
curium	Cm	96	–	rhodium	Rh	45	102.905
dysprosium	Dy	66	162.50	rubidium	Rb	37	85.47
einsteinium	Es	99	–	ruthenium	Ru	44	101.07
erbium	Er	68	167.26	samarium	Sm	62	150.35
europium	Eu	63	151.96	scandium	Sc	21	44.956
fermium	Fm	100	–	selenium	Se	34	78.96
fluorine	F	9	18.9984	silicon	Si	14	28.086
francium	Fr	87	–	silver	Ag	47	107.868
gadolinium	Gd	64	157.25	sodium	Na	11	22.9898
gallium	Ga	31	69.72	strontium	Sr	38	87.62
germanium	Ge	32	72.59	sulfur	S	16	32.064
gold	Au	79	196.967	tantalum	Ta	73	180.948
hafnium	Hf	72	178.49	technetium	Tc	43	–
helium	He	2	4.0026	tellurium	Te	52	127.60
holmium	Ho	67	164.930	terbium	Tb	65	158.924
hydrogen	H	1	1.00797	thallium	Tl	81	204.37
indium	In	49	114.82	thorium	Th	90	232.038
iodine	I	53	126.9044	thulium	Tm	69	168.934
iridium	Ir	77	192.2	tin	Sn	50	118.69
iron	Fe	26	55.847	titanium	Ti	22	47.90
krypton	Kr	36	83.80	tungsten	W	74	183.85
lanthanum	La	57	138.91	uranium	U	92	238.03
lead	Pb	82	207.19	vanadium	V	23	50.942
lithium	Li	3	6.939	xenon	Xe	54	131.30
lutetium	Lu	71	174.97	ytterbium	Yb	70	173.04
magnesium	Mg	12	24.312	yttrium	Y	39	88.905
manganese	Mn	25	54.9380	zinc	Zn	30	65.37
mendelevium	Md	101	–	zirconium	Zr	40	91.22

ENTHALPY-ENTROPY DIAGRAM

ENTHALPY-ENTROPY (MOLLIER) DIAGRAM FOR STEAM
(English units)

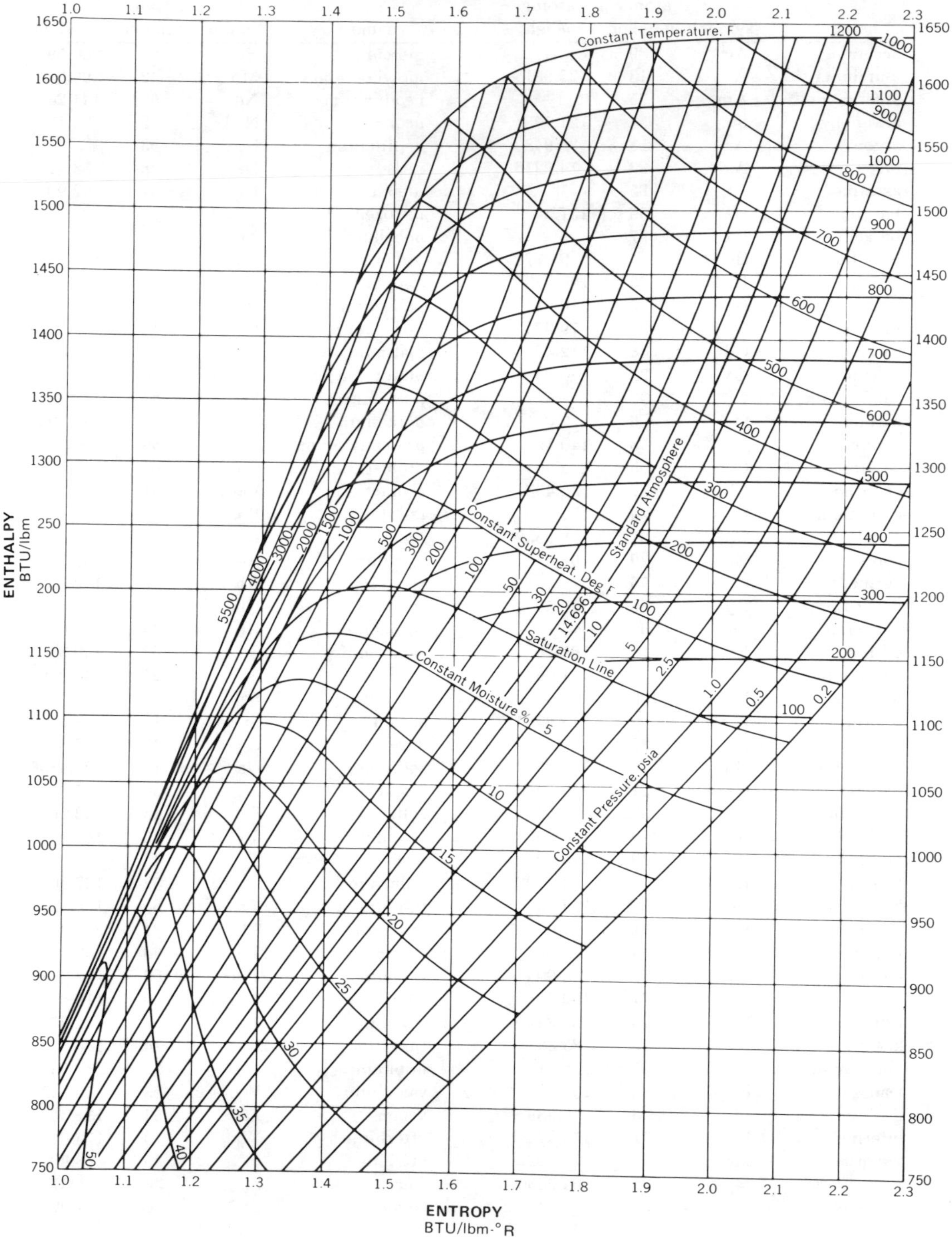